International Conference on Trends in Spintronics and Nanomagnetism 2010

(TSN 2010)

Journal of Physics: Conference Series Volume 292

Lecce, Italy
23 – 27 May 2010

ISBN: 978-1-61782-767-9
ISSN: 1742-6588

Printed from e-media with permission by:

Curran Associates, Inc.
57 Morehouse Lane
Red Hook, NY 12571

CURRAN ASSOCIATES INC. proceedings.com

Some format issues inherent in the e-media version may also appear in this print version.

Printed by Curran Associates, Inc. (2011)

Institute of Physics
Dirac House, Temple Back
Bristol BS1 6BE UK

Phone: 44 1 17 929 7481
Fax: 44 1 17 920 0979

techtracking@iop.org

Additional copies of this publication are available from:

Curran Associates, Inc.
57 Morehouse Lane
Red Hook, NY 12571 USA
Phone: 845-758-0400
Fax: 845-758-2634
Email: curran@proceedings.com
Web: www.proceedings.com

TABLE OF CONTENTS

PAPERS

PREFACE

The International Conference "Trends in Spintronics and Nanomagnetism" (TSN-2010), was organized by partners of the EU-project SpiDME and held in the historical city of Lecce, Italy from 23–27 May 2010, at the Ecotekne Campus, University of Salento. The conference provided an international forum to discuss recent progress and future trends in the field. In particular, the aim was to bring together the community of more conventional spin-transport, with that of molecular and nano-magnetism. The main topics of TSN-2010 were:

MATERIALS

– Molecular Magnets

– Magnetic nanoparticles and nanowires

– Magnetic semiconductors

– Multiferroics and transition metal oxides

SPIN-PHYSICS AND THEORY

– Spin injection

– Domain walls, spin torque and vortex dynamics

– Numerical modeling of organic nanomagnetism

APPLICATIONS

– Magnetic Multilayers

– Spin-photonics

– Molecular and nano-spintronics

– Spin-based quantum computation

– Magnetism for sensing and nanomedicine

ADVANCES IN CHARACTERIZATION

– Magneto-optical characterization and spin manipulation

– Intrinsic spin transport mechanism in organics

– Organometallic molecules on surfaces

– Single molecular magnets on surfaces

– Nanoscale characterization and spin-sensitive SPM

The scientific programme started on Monday 24 May and ended on Thursday 27 May. The Nobel Laureate A Fert attended the conference giving a plenary talk and the programme also featured invited presentations by (in alphabetical order): M Aeschlimann, M Affronte, N Atodiresei, P A Bobbert, A Dediu, N Kioussis, L W Molenkamp, J Moodera, V Prigodin, M Ruben, R Sessoli, R Tan, and H Wende.

TSN2010 had 150 attendees who came from around the globe to present their latest research in 100 oral presentations. Contributed talks were selected by the program committee, composed of Giuseppe Maruccio, Ross Rinaldi, Valentina Arima, Fabio Della Sala, Maurizio Martino (Universitá del Salento, NNL Institute Nanoscience-CNR, Lecce, Italy), Stefano Sanvito (Trinity College Dublin, Ireland), Germar Hoffmann and Roland Wiesendanger (Institute for Applied Physics, University of Hamburg, Germany), and Alan Rowan (NSRIM Institute Radboud University Nijmegen, The Netherlands). This group also acted as the Publication Committee and managed all the submitted papers that were reviewed by expert referees in order to meet the standards of the *Journal of Physics: Conference Series*.

Nobel Laureate A Fert with some members of the organizing committee.

The conference would not have been possible without the support from the local organizing committee at the University of Salento and NNL Institute Nanoscience-CNR, including Anna Paola Caricato, Luigi Martina and the Conference Secretaries Maria Concetta Gerardi, Adriana Amato, and Gabriella Zammillo. We are grateful for the technical assistance of Michele Linciano, Antonio Guerrieri, Carmine Mangia, Luciano Carluccio, and Tommaso Moscara e Francesco Sabetta. We also gratefully acknowledge Serena Chiriacó, Anna Grazia Mondeduro and Massimo Corrado who helped to run the conference.

The conference was made possible by the financial support from the European Commission through the SpiDME project (EU-FP6-029002), the Italian Ministry for Foreign Affairs, the University of Lecce and its Department of Physics, and all of the sponsors (Lot Oriel, Attocube, Schaefer, Cryogenic Ltd, Oxford Instruments, MTI Corporation, Cantele, Monte dei Paschi di Siena).

Conference Chair and Co-Chairs
Giuseppe Maruccio
Stefano Sanvito
Germar Hoffmann
Roland Wiesendanger
Alan Rowan

All papers published in this volume of *Journal of Physics: Conference Series* have been peer reviewed through processes administered by the proceedings Editors. Reviews were conducted by expert referees to the professional and scientific standards expected of a proceedings journal published by IOP Publishing.

Journal of Physics: Conference Series **292** (2011) 012001 doi:10.1088/1742-6596/292/1/012001

New advances in organic spintronics

V N Prigodin[1,2], J W Yoo[1,3], H W Jang[4], C Kao[3], C B Eom[4], and A J Epstein[1,3]

[1]Department of Physics, Ohio State University, Columbus, Ohio 43210-1117, USA
[2]Ioffe Institute, 194021 St. Petersburg, Russia
[3]Department of Chemistry, Ohio State University, Columbus, Ohio 43210-1137, USA
[4]Department of Materials Science and Engineering, University of Wisconsin–Madison, Madison, Wisconsin 53706, USA

E-mail: prigodin.1@osu.edu

Abstract. The basic components of spintronic devices are spin polarized ferromagnets and spin transporting non-magnetic spacers. Exploiting carbon-based materials for these components promises to extend functionality of information storage and processing as well as to improve device integration and fabrication. Here we present the magnetoresistance of organic semiconductor rubrene ($C_{42}H_{28}$) used as a spacer in $La_{2/3}Sr_{1/3}MnO_3$ (LSMO)/organic semiconductor (OSC)/Fe heterojunctions. Efficient spin polarized tunneling through the thin layer of rubrene spacer (5 nm) was observed. As the thickness of rubrene layers is increased, device current is strongly limited by carrier injection resulting in strong temperature dependent device resistance. The carrier injection is described with thermionic field emission at the metal/OSC interface. As a next step toward organic spintronics we used an organic based magnet vanadium-tetracyanoethylene ($V(TCNE)_x$, x~2) in tandem with LSMO in a spin- valve with 5 nm rubrene spacer. $V(TCNE)_x$ is the earliest developed room temperature molecule-based magnet (T_c ~ 400 K). Due to strong on-site Coulomb interaction and weak intermolecule overlapping their magnetic state can be described with a model of half-semiconductor in which valence and conduction bands are spin polarized. The magnetoresistance data for bulk $V(TCNE)_x$ is in agreement with the model of spin polarized valence and conduction bands. We demonstrated that an organic-based magnetic semiconductor $V(TCNE)_x$ functions very well as an electron spin polarizer in the standard spintronic device geometry.

1. Introduction

An emerging field of "spintronics" based on the control of electron "spin" degree of freedom introduces a number of new phenomena and applications [1]. The principal spintronic device, a "spin valve", consists of soft and hard magnetic layers separated by a non-magnetic spacer. Using of organic materials for these components potentially enables to improve functionality and processability [2, 3]. Here we present our recent progress in this organic spintronics.

The application of organic semiconductors (OSCs) as a spacer was motivated by anticipated long spin lifetime in OSCs. However, efficient spin polarized carrier injection and transport in OSC films are rather complicated [4]. Definitely, the spin valve effect was measured in thin layers of OSC (< 10 nm) as tunneling magnetoresistance (TMR) [5, 6]. But the previous reports [7, 8] on giant magnetoresistance (GMR) with thick OSC layers (> 100 nm) have been very controversial [9-11]. The low and temperature/voltage independent resistance as well as wrong sign of magnetoresistance for those devices suggest that the charge transport in the spacer represents tunneling through a random network of metallic (Co) nanoparticles. The OSC plays rather a role of supporting dielectric matrix for those metallic granular networks.

Here we describe real GMR in OSC films due to injection and transport through the OSC channel [4]. We used rubrene as a prototype of OSC because rubrene consists of only hydrogens and carbons and has the highest mobility among OSCs and a low energy gap ~ 2.3 eV (for Alq3 ~ 2.8 eV). Varying thickness of rubrene film from 5 to 50 nm we observed the crossover from TMR to GMR. In the GMR regime the rubrene devices show strong exponential temperature and electric filed dependence of the current and sign of GMR remains the same as TMR one. The maximal thickness of rubrene film for which we detect MR is 30 nm at temperature less than 100 K.

Another advance we introduced was the use of a molecular based magnet $V(TCNE)_x$ as spin injector/detector for a spin valve [12]. It is widely spread that the spin transfer between semiconductor spacers and magnetic contacts can be improved by using magnetic semiconductors typically obtained by doping conventional semiconductors with transition metals [1]. However the application of such diluted magnets is restricted because of their low Curie temperature. The organic $V(TCNE)_x$ is a semiconductor with T_c ~ 400 K [13]. Here we report that $V(TCNE)_x$ can be used as spin injector/detector in a spin valve with a 5 nm rubrene film as spacer. The device demonstrates excellent performance and its operation at high temperatures is limited to spin polarization of LSMO which is used as the second magnetic contact.

2. Giant Magnetoresistance with OSC, Rubrene

Fig. 1a illustrates the schematics of our LSMO(50nm)/LAO(1.2nm)/Rubrene/Fe(30nm) heterojunction with rubrene ($C_{42}H_{28}$) film as a spacer. The epitaxial 50 nm of LSMO was grown by pulse laser deposition. Thin layer (1.2 nm) of $LaAlO_3$ (LAO) was employed to improve interfacial quality between LSMO and rubrene. The thickness of the rubrene layer in our devices was varied from the TMR regime ($d \leq 5$ nm) to the giant magnetoresistance limit ($d \geq 20$ nm). The temperature dependence of MR curves for 5 nm rubrene device at bias voltage 10 mV is presented in Fig. 1b. Device exhibits positive MR over all biases and temperatures and the steps of MR show excellent correspondence with each other as well as the coercivity of Fe (30 nm) and LSMO (50 nm) electrodes.

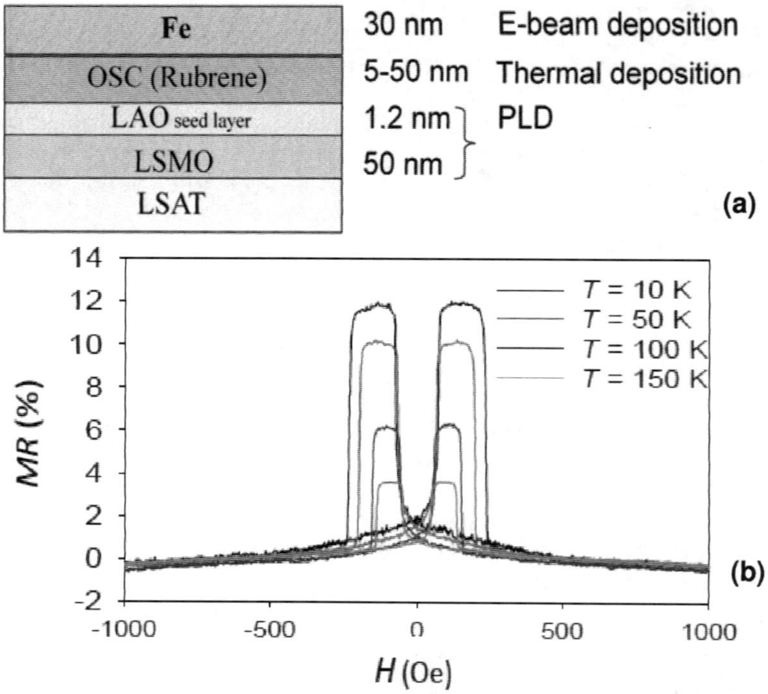

Figure 1 **(a)** Schematic view on the device structure. **(b)** Tunnel magnetoresistance of LSMO(50 nm)/LAO(5 nm)/Ru(5nm) /Fe(30 nm) heterojunction. Reproduced in part with permission from Yoo et al., Phys. Rev. B 80, 205207 (2009) Copyright © 2009 American Physical Society.

As the thickness of rubrene is increased, the device current strongly relies on carrier injection resulting in strong temperature and voltage dependent device resistance [4]. Fig. 2a displays T dependent device resistivity for the 5, 20, 50 nm rubrene devices and 50 nm LSMO film at bias 0.1 V. Strong nonlinear I-Vcurves are shown in Fig. 2b for 5, 20, 30, and 50 nm Ru devices.

The strong T and V_b dependent current for devices with space thickness less than 100 nm originates mainly from activation at the metal/OSC interface barriers rather than bulk resistance. To explain the voltage and temperature dependence of current in our organic-based spin valve devices, we use the theoretical model developed by Kiveris and Pipinys [14]. This model describes carrier injection at the metal/semiconductor interface by equally putting multiphonon activation and field emission of electrons from defect states at the interfaces to the conduction band of semiconductors. The phonon assisted tunneling rate of electron under the electric field at the metal/semiconductor interface is as follow [14, 15]

$$W_T = eE(8m^*\varepsilon_T)^{-1/2}(1+\gamma^2)^{1/2}-\gamma]^{1/2}[1+\gamma^2]^{-1/4}$$
$$\times\exp\{-(32m^*\varepsilon_T^3)^{1/2}(3eE\hbar)^{-1}[(1+\gamma^2)^{1/2}-\gamma]^2[(1+\gamma^2)^{1/2}+\gamma]^{1/2}\}, \quad (1)$$

where $\gamma = (2m^*)^{1/2}\,\Gamma^2\,(8\hbar eE)^{-1}\varepsilon_T^{-1/2}$. Here ε_T is the energetic depth of defect states from the conduction band of semiconductor, $E = V_b/d$ is the applied electric field, $\Gamma^2 = 8a(\hbar\omega)^2(2n+1)$ is the width of the defect states broadened by the interaction with optical phonons, $n = 1/[\exp(\hbar\omega/k_BT) - 1]$, and a is the electron-phonon coupling constant ($a = \Gamma_0/8(\hbar\omega)^2$).

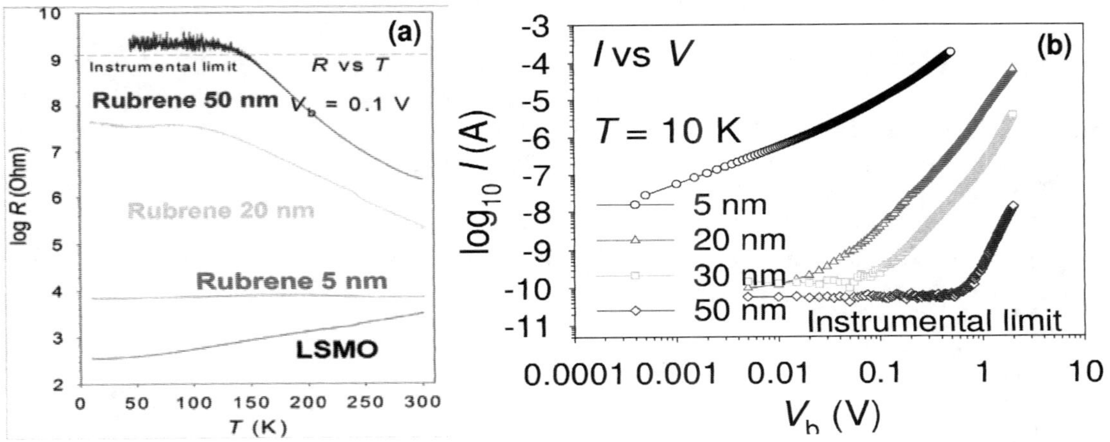

Figure 2 (a). T-dependence of resistance of 5, 20, 50 nm rubrene devices and 50 nm LSMO film at bias 0.1 V. **(b)** I-Vcurves for 2, 20, 30 and 50 nm rubrene devices at T = 10 K. Reproduced in part with permission from Yoo et al., Phys. Rev. B 80, 205207 (2009) Copyright © 2009 American Physical Society.

The dashed lines in Fig. 3 show fits to 20 nm rubrene device current with Eqn. (1). Qualitatively good fitting can be achieved over all $T$and V_b with fixed parameter of a = 1.4, ħω = 0.034 eV, m^* = 1.2 m_e and ε_T = 0.678 eV (for V_b = 0.1 V), and 0.660 eV (for V_b = 0.1 V). The results show that 20 nm rubrene device currents crucially depend on the carrier injection. The phonon assisted carrier injection start to increase dramatically above 100 K. Below 100 K the temperature independent but strongly voltage dependent elastic tunneling dominates.

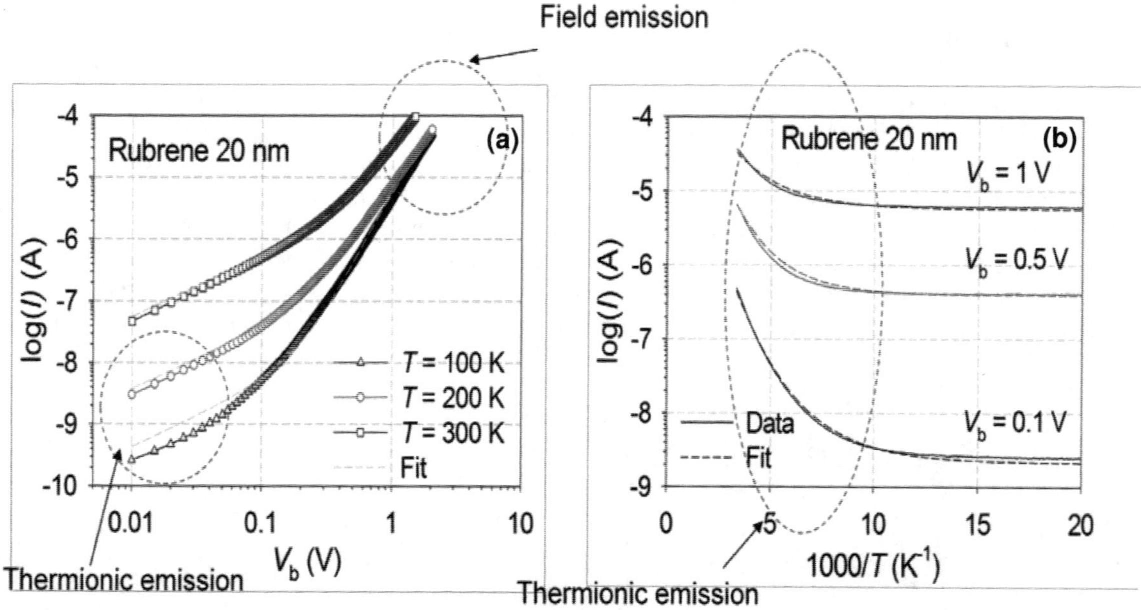

Figure 3. Temperature and bias dependencies of the current for 20 nm rubrene device. The dashed lines show the fit to the phonon-assisted tunneling model. Reproduced in part with permission from Yoo et al., Phys. Rev. B 80, 205207 (2009) Copyright © 2009 American Physical Society.

As the thickness of rubrene layer is increased further ($d > 30$ nm), the effects of carrier transport in the bulk on the device current become stronger. As it was suggested by Ruden and Smith [15], in order to transmit spin polarized carriers through the thick amorphous OSC channel, applying high bias is essential to promote field emission and support higher drift mobility (e.g., $\mu = \mu_0 \exp(E/E_0)^{1/2}$). However, the conductivity mismatch problem could be more serious in electrical detection of MR [17] as d increases further. The MR of 20 nm rubrene film is shown on Fig. 4a at different temperature. Fig. 4b shows the MR for spin valves with diffferent thckness of rubrene spacer. No any MR signal was detected for thickness larger than 30 nm.

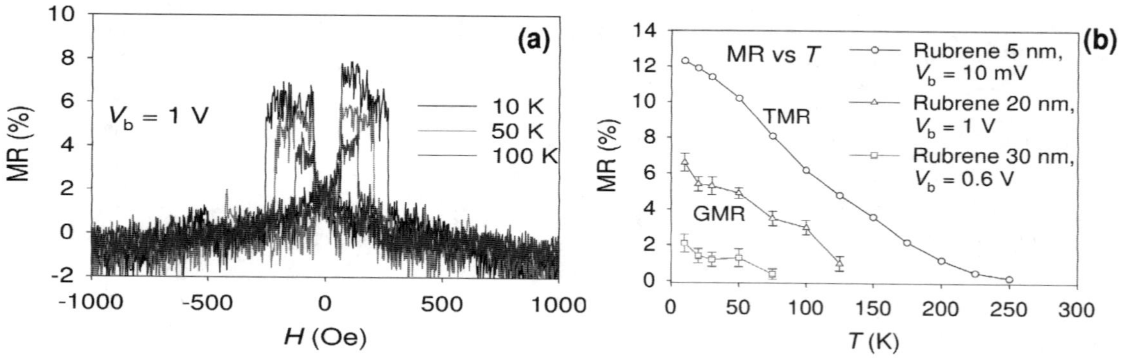

Figure 4 (a). MR of 20 nm rubrene device for different temperatures and bias 1 V. **(b)** Temperature dependence of MR for 5, 20, and 30 nm rubrene devices. Reproduced in part with permission from Yoo et al., Phys. Rev. B 80, 205207 (2009) Copyright © 2009 American Physical Society.

3. Molecular based Magnetic Semiconductor, $V(TCNE)_2$

Vanadium-tetracyanoethylene ($V(TCNE)_x$, x~2) represents a room-temperature molecule-based ferrimagnet with ordering temperature T_c ~ 400 K and a small coercive field H_c = 4.5 G. $V[TCNE]_2$ is a semiconductor with conductivity ~10^{-4} S/cm and gap ~ 0.5 eV at 300 K. It can be prepared at 40 °C using chemical vapor deposition [18].

Nature of magnetic state and semiconductor charge transport of $V(TCNE)_2$ can be understood within the model of half semiconductor [13]. From average stiochiometry $V(TCNE)_2$ and because the $[TCNE]^-$ is the radical anion, it can conclude that the vanadium is in the V^{II} oxidation state [19]. The spins of three electrons at the 3d level of V^{II} are parallel with a total spin of 3/2. Fig. 5a represents a one-dimensional "cartoon" for the spatial distribution of energy levels in $V(TCNE)_2$. The orientation of electronic spins in the π^* state is opposite of the 3/2 spin of V^{II} due to the large antiferromagnetic exchange coupling, J.

The relevant model for description of electronic state on TCNEs is the Hubbard model with nearly half filling π^* band [13]. According to this model due to the Coulomb repulsion U_c between electron sitting on the same site the initial π^* band is split into two non overlapping subbands provided $U_c >> t$, where t is the electronic transfer integral between neighboring TCNEs. The spins of electrons in the lower filled Hubbard subband should have antiferromagnetic order, but the corresponding exchange constant, $J' = 2t^2/U_c$, is much less than the antiferromagnetic exchange constant, J, between 3/2 spins of V^{II}s and spin ½ of $[TCNE]^-$. Therefore, in the magnetic phase the electronic spins of lower subband are parallel and their orientation is opposite to the spin of V^{II}. According to the exclusion principle, the upper empty subband may be occupied by electrons with spins polarized in the direction which is opposite to the lower subband, i.e., along the direction of the V^{II} spins.

The one-site Coulomb repulsion U_c or the gap between π^* and $\pi^* + U_c$ subbands in analogy with known organic materials is estimated to be 2 eV [20]. According to Tengstedt et al. [20-21], the $3d$ level of V is located inside the Hubbard gap. Therefore, the temperature dependence of conductivity is determined by thermal activation of electrons from $3d$ levels of Vs to $\pi^* + U_c$ levels of TCNEs (see Fig. 5a). The corresponding activation energy, E, besides the energy separations between the two levels, $E_0 = E(\pi^* + U_c) - E(3d)$ ~ 0.5 eV, includes the change of magnetic energy associated with this electronic transition.

Taking into account the opposite spin polarizations of π^* subbands and their antiferromagnetic coupling with V^{II} spins, the magnetic contribution to the activation energy, ΔE, is approximated as $\Delta E - \Delta E_0 = -4J\langle S\rangle\langle\sigma\rangle$ [3, 22], where $\langle S\rangle$ is the spin 3/2 polarization of V^{II} and $\langle\sigma\rangle$ is the spin 1/2 polarization of the lower π^* subband. Applying the magnetic field changes the spin polarization, and hence the activation energy. In this way, the magnetoresistance can be estimated from the relation

$$MR = (-4J/k_BT)[\langle S(H)\rangle\langle\sigma(H)\rangle - \langle S(0)\rangle\langle\sigma(0)\rangle]. \quad (2)$$

In the paramagnetic phase the spin polarizations are induced by an external magnetic field and become noticeable only upon approaching T_c. Within the mean field theory [12] $\langle S\rangle \sim -\langle\sigma\rangle \sim \chi h$, where χ is the Curie-Weiss susceptibility per spin, $\chi \sim 1/\delta$, with $\delta = (T - T_c)/T_c$ being the "distance" from T_c (k_BT_c ~3.2 J). Here $h = \mu_B gH/(k_BT_c) << 1$ and $h^{2/3} << \delta << 1$. As a result, the gap widens with the applied field and $MR \sim (h/\delta)^2$.

In the critical region, $|\delta| << h^{2/3}$, the induced polarization obeys the scaling law $\langle S\rangle \sim -\langle\sigma\rangle \sim h^{1/3}$ and MR ~ $h^{2/3}$. Below T_c ($|\delta| >> h^{2/3}$, $\delta < 0$), there is the spontaneous spin polarization $\langle S\rangle \sim -\langle s\rangle \sim |\delta|^{1/2}$, with field-dependent corrections ~ χh, where $\chi \sim 1/|\delta|$. Therefore, in the ferrimagnetic phase the MR is proportional to a magnetic field, MR ~ $h/|\delta|$.

To check these predictions on the H and T dependencies of MR, the T_c was decreased by exposition of a sample on air. Fig. 5b shows up the magnetic field dependence of magnetoresistance below and above T_c = 240 K. Linear and quadratic dependence of magnetoresistance on a magnetic field clearly can be identified. As a function of temperature, the MR has its maximum near T_c. For H = 6 kG and T_c = 230 K the model predicts the MR maximum ~ $h^{2/3}$ ~ 2%, which is of the order of the observed value.

Journal of Physics: Conference Series **292** (2011) 012001 doi:10.1088/1742-6596/292/1/012001

Thus, the experimental results for resistance and MR of organic-based magnet $V(TCNE)_x$ support the existence of spin-polarized subbands. This is a very desirable property for utilizing organic-based magnetic semiconductors in spintronic applications.

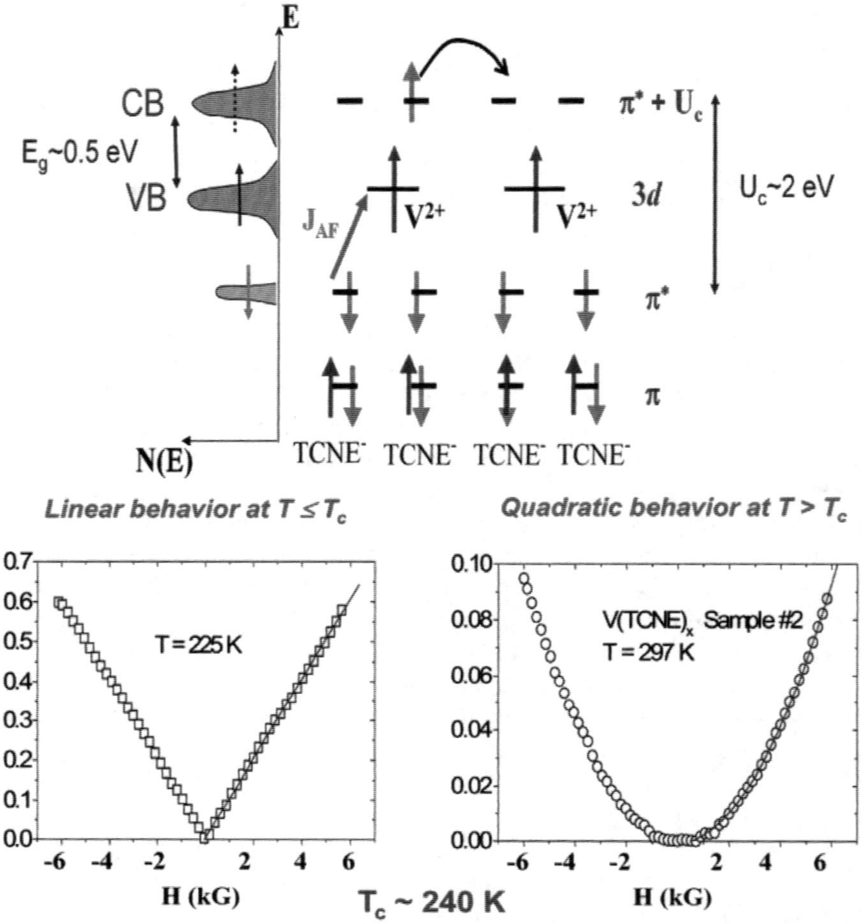

Figure 5 (a). Energy diagram for $V(TCNE)_2$. (Reproduced with permission from Prigodin et al., Adv. Mater. 14, 1230 [2002]. Copyright @ 2002 John Wiley and Sons.) **(b)** The magnetoresistance of $V(TCNE)_2$ as a function of magnetic field above and below T_c. Reproduced with permission from Prigodin et al., Adv. Mater. 14, 1230 (2002). Copyright © 2002 John Wiley and Sons.

4. Organic based Spin Valve V(TCNE)$_x$(500 nm)/Rubrene(5 nm)/LAO(1.2 nm)/LSMO(80 nm)

Here we demonstrate that the organic-based magnetic semiconductor $V(TCNE)_x$ functions well as the ferromagnetic layer in a hybrid magnetic tunnel junction device [12]. The rubrene thin film was used for organic semiconducting barrier as it was shown to be highly efficient for spin polarized tunneling. Fig. 6a shows typical magnetoresistance curves for in-plane magnetic field in the hybrid magnetic tunnel junction $V(TCNE)_x$(500 nm)/Rubrene(5 nm)/LAO(1.2 nm)/LSMO(80 nm) at 100 K with an applied bias of 0.5 V. The magnetization curves for individual magnetic layers of $V(TCNE)_x$ and LSMO films were recorded by SQUID magnetometry and are overlaid on the same plot of Fig. 6a. Blue line and symbols (△) represent the data recorded during positive sweep of magnetic field. Red line and symbols (▽) represent the data collected during negative field sweeping.

With sweeping of the magnetic field, the device resistance becomes higher when the magnetizations of two magnetic layers become antiparallel. The magnetoresistance curves well correspond to the magnetization steps of both V(TCNE)$_x$ and LSMO films and clearly represent the shape of hysteresis curves of each individual magnetic layer. This unambiguously demonstrates spin filtering through a V(TCNE)$_x$ magnetic semiconductor film, wherein spin polarized carriers tunnel through the hybrid rubrene/LAO barrier while effectively conserving their spin polarization.

The MR as a function of temperature (Fig. 6b) shows a maximum at ~ 100 K. The decrease of MR, as temperature increases above 100 K, is attributed to the decrease of surface spin polarization of the LSMO film. This suggests that MR at higher temperature can be achieved in magnetic tunnel junctions with organic-based magnet V(TCNE)$_x$ in tandem with other ferromagnets. The MR of our devices decreases as temperature decreases below 100 K because the semiconductor bulk resistance of V(TCNE)$_x$ film starts to dominate the total device resistance as T decreases below 100 K [12].

The bias dependence of MR in our hybrid magnetic tunnel junction is similar to the behavior of tunnelling magnetoresistance in typical inorganic magnetic tunnel junctions. At lowest bias, data are noisier as the device resistance becomes high at low T and low V_b. The observed bias-induced suppression of MR by the application of high-voltage could be also related with the bulk resistance of V(TCNE)$_x$ film. At low bias, the barrier resistance controls the device currents. As the applied bias increases, the bulk resistance of V(TCNE)$_x$ starts to control the total device resistance, which reduces the effectiveness of electrical detection of spin dependent scattering at the interface.

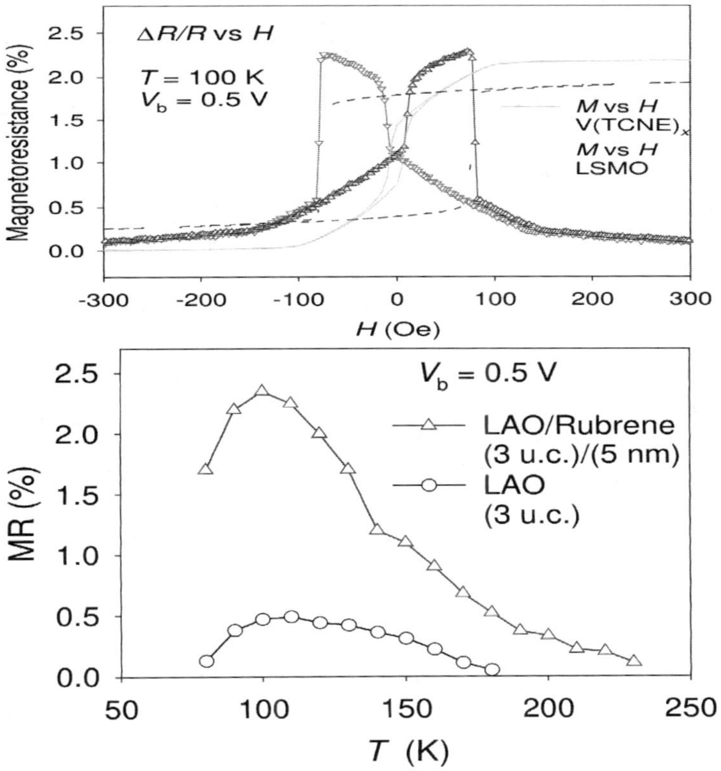

Figure 6 (a). MR of organic based spin valve V(TCNE)$_x$(500 nm)/Rubrene(5 nm)/LAO(1.2 nm)/LSMO(80 nm) and magnetization loops for LSMO (dark) and V(TCNE)$_2$. **(b)** MR of organic based spin valve as a function of temperature. Reproduced with permission from Yoo et al., Nat. Mat., 9, 638 (2010) Copyright © 2010 Macmillan Publishers Limited.

5. Conclusion

In summary, we presented the GMR with an OSC (Rubrene) spacer. In contrast to the previous reports the GMR for our devices shows strong V_b and T dependence as a result of the activation processes of carrier injection and transport in amorphous OSC layer. We explained the carrier injection and transport into/in OSC and their impact on the GMR in OSC-based spin valves in terms of phonon-assisted tunneling and hopping in the presence of strong electric field.

We have shown that the magnetotransport of organic based magnet $V(TCNE)_2$ can be explained in terms of discrete spin polarized bands with hopping mechanism transport. Spin valve used $V(TCNE)_2$ as one of magnet confirms the presence of these spin polarization bands.

6. Acknowledgement

The present work was supported by DOE (Grants No. DE-FG02-01ER45931, DE-FG02-86ER45271), NSF (Grant No. DMR-0805220), AFOSR (Grant No. FA9550-06-1-0175).

References

[1] Wolf S A, Awschalom D D, Buhrman R A, Daughton J M, von Molnár S, Roukes M L, Chtchelkanova A Y and. Treger D. M 2001 *Science* **294** 1488
[2] Epstein A J and Prigodin V N 2000 *US Patent* 6 621 100
[3] Sanvito S and. Rocha A R 2006 *J. Comp. Theor. Nanosci.* **3** 624
[4] Yoo J-W, Jang H W, Prigodin V N, Kao C, Eom C B and Epstein A J 2009 *Phys. Rev.* **B 80** 205207
[5] Moodera J S, Kinder L R, Wong T M and R. Meservey, 1995 *Phys. Rev. Lett.* **74** 3273
[6] Shim J H, Raman K V, Park Y J, Santos T S, Miao G X, Satpati B and Moodera J S 2008 *Phys. Rev. Lett.* **100** 226603
[7] Dediu V, Murgia M, Matacotta F C, Taliani C and Barbanera S 2002 *Solid State Commun.* **122** 181
[8] Xiong Z H, Wu D, Vardeny Z V and Shi J 2004 *Nature* (London) **427** 821
[9] Santos T S, Lee J S, Migdal P, Lekshmi I C, Satpati B and Moodera J S 2007 *Phys. Rev. Lett.* **98** 016601
[10] Vinzelberg H, Schumann J, Elefant D, Gangineni R. B, Tomas J and Büchner B 2008 *J. Appl. Phys.* **103** 093720
[11] Jiang J S, Pearson J E and Bader S D 2008 *Phys. Rev.* B **77** 035303
[12] Yoo J-W, Chen C-Y, Jang H W, Bark C W, Prigodin V N, Eom C B and Epstein A J 2010 *Nature Materials* **9** 638
[13] Prigodin V N, Raju N P, Pokhodnya K I, Miller J S and Epstein A J 2002 *Adv. Mater.* **14** 1230
[14] Kiveris A, Kudzmauskas S and Pipinys P 1976 *Phys. Stat. Sol.* **37** 321
[15] Pipinys P and Kiveris A 2005 *J. Phys.: Condens. Mater.* **17** 4147
[16] Ruden P P and Smith D L 2004 *J. Appl. Phys.* **95** 4898
[17] Schmidt G, Ferrand D, Molenkamp L. W, Filip A T and van Wees B J 2000 *Phys. Rev.* B **62** R4790
[18] Pokhodnya K I, Epstein A J and Miller J S 2000 *Adv. Mater.* **12** 410
[19] Miller J S and Epstein A J 1994 *Angew. Chem. Int. Ed. Engl.* **33** 385
[20] Tengstedt C, Unge M, de Jong M P, Stafstrom S, Salaneck W R and Fahlman M 2004 *Phys. Rev.* B **69** 165208
[21] Tengstedt C, de Jong M P, Kanciurzewska A, Carlegrim E and Fahlman M 2006 *Phys. Rev. Lett.* **96** 057209
[22] Raju N P, Prigodin V N, Pokhodnya K I, Miller J S and Epstein A J 2010 *Synth. Met.* **160** 307

Journal of Physics: Conference Series **292** (2011) 012002 doi:10.1088/1742-6596/292/1/012002

Nano-electronics and spintronics with nanoparticles

S Karmakar[1], S Kumar[1], R Rinaldi[1,2] and G Maruccio[1,3]

[1]NNL Istituto Nanoscienze-CNR, Lecce, Italy
[2]Department of Innovation Engineering, University of Salento, Via Arnesano, 73100, Lecce, Italy
[3]Physics Department, University of Salento, Via Arnesano, 73100, Lecce, Italy

Email: shilpi.karmakar@unisalento.it , giuseppe.maruccio@unisalento.it

Abstract. We review the current research on nanodevices with nanoparticles which present unique challenges in both the realization of well-controlled interfaces at the nanoscale and the ability to adequately characterize their electrical properties. In particular, we discuss the fabrication and electrical characterization of such nanodevices with special attention to devices based on metal and magnetic nanoparticles.

1. Introduction

Over the past decades, progress in miniaturization of electronic devices has played a powerful role in the continuous technological development, increasing the performance and speed of the information exchange. Advances in nanolithographic fabrication techniques and chemical methods like self-assembly made possible to scale down the devices to few tens of nanometers or less [1-3]. This has already made a great impact on the performance of traditional semiconductor circuits and opened up new possibilities based upon quantum effects. The same has led to the birth of a new field called single electronics where the exploitation of charging effects in tunnel junctions with nanoscale islands allows transport/control of individual charge carriers [4]. In metals these carriers are electrons while in the case of superconducting materials at low temperature, they can be Cooper pairs.
In this review, we summarize recent studies of nanoelectronics with metallic and magnetic nanoparticles (NPs) with a view towards future applications and spintronics. Vertical electron transport has been studied extensively to elucidate functionalities like non-linear I–V characteristics and nanoscale ohmic contacts. The most prominent amount of the published work relies heavily on scanning probe microscopy [5-16]. However for practical applications, the deposition of clusters/single nanoparticles within nanojunctions fabricated on insulating surfaces is required. Patterning or positioning can be accomplished through self-assembly of the nanocrystals or could either be achieved via electrostatic interactions at the cluster–substrate interface [17, 18]. The ability to functionalize a nanoparticle network with linking molecules provides appealing low cost design flexibility.

2. Single Electron Transistors

One of the most promising nanoelectronic devices is the single-electron transistor (SET) [4], where single-electron charging effects are used to precisely control the charging of a conducting island by individual electrons. Electrons are transferred to the island from source and drain terminals by tunnelling across potential barriers and the charging process can be controlled by a gate terminal (Fig. 1). The SET has the advantages of very low power consumption and better immunity from statistical charge fluctuation, compared to conventional devices. In the past two decades SETs made from metallic nanoparticles have attracted extensive research interest for their potential applications in future nanoelectronics viz. integrated circuits and supersensitive electrometers [19-23].
The practical application of SET requires room temperature operation and silicon process compatibility. For room temperature operation, the single-electron charging energy of the island, $E_C = e^2/2C_{TOT}$ (C_{TOT} is the total island capacitance and e is the elementary electronic charge) must be large compared to the thermal energy $\sim k_BT \approx 26$ meV (k_B is Boltzmann's constant and temperature T= 300 K). Therefore, the value of the C_{TOT} must be ~1 aF or less and in practice this implies that the charging island must be <10 nm in size. In addition, electrons must be localized on the island, which requires a large tunnel barrier resistance as compared to the quantum resistance R_Q~26 kΩ at the operating temperature. The use of very small metallic islands also improves the charge sensitivity of the electrometers beyond the operational temperature range.
For SET fabrication usually chemically synthesized colloidal nanocrystals are preferred since they have smaller and more uniform sizes, while another advantage is their potential for controlled assembly. However, integrating individual nanoparticles into devices and gating them effectively is extremely challenging. This is due to the challenges in (i) fabricating nanometer-spaced electrodes and (ii) the precise placement of the nanoparticles between the electrodes. Several techniques have been developed for the fabrication of SETs using metal nanoparticles.

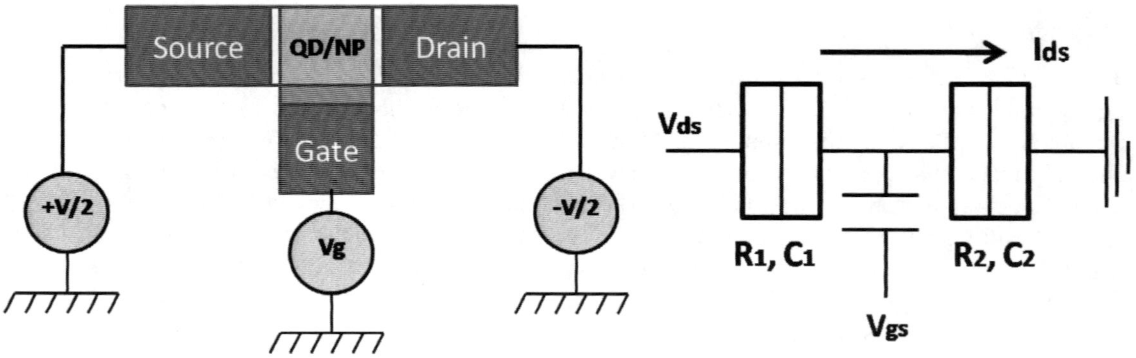

Figure 1. (a) Schematics design of a single nanoparticle between two metal electrodes. (b) Equivalent electrical circuit of a single electron transistor.

3. Nanojunction fabrication

There are mainly two approaches for connecting nanocrystals between electrodes. The first one is to make top-contact junctions, which includes scanning tunnelling microscopy (STM) [6-14] and conducting atomic force microscopy (AFM)) [15, 16]. Although very useful for fundamental investigations, this strategy is, however, far from practical applications and makes difficult to introduce a third gate electrode. The second approach which overcomes these limitations is to fabricate devices based on nanogap electrodes [17], a strategy potentially downscalable opening the way to realize integrated circuits. In the last two decades, prominent efforts has been made to develop novel fabricating techniques for nanodevice realization. Several effective and novel methods of fabricating nanogap electrodes with controlled spacing have been reported, including electron-beam lithography [24-26], electromigration [27-29], mechanical break junctions [30, 31], electrochemical plating [32], focused ion beam lithography [33], shadow mask evaporation [34], scanning probe and atomic force microscopy lithography [35], on-wire lithography [36, 37], etc. Different methods are frequently combined as well to obtain a desired nanocircuit arrangement.

3.1 Mechanical Controllable Break Junctions (MCBJ)

This method was first adopted by Reed et al. [31] to fabricate nanogap electrodes, which yielded electrodes with a separation of few nanometers. In this process, a notched metallic wire is generally glued to an elastic substrate (Fig.2 (a)), which serves as the bending beam. The substrate is bent by pushing its centre with a driving rod and consequently the notched wire is fractured, after which an adjustable tunnelling gap can be established. The surfaces can be brought together again, and the distance can be controlled by a piezoelectric element. The breaking process is mostly conducted under low temperature and high vacuum conditions, and this guarantees two atomically clean surfaces. The MCBJ technique is very stable and the contact separation can be continuously adjusted under the control of an ultrafine piezoelectric component without polluting the junction [38-40]. Although the MCBJ method is really useful for fundamental investigation, it is not facile to fabricate highly integrated nanojunction arrays because of the constraint of the piezoelectric components, and it is difficult to build devices with three or even more electrodes or electrodes with larger separation.

3.2 Electrochemical Deposition for Nanogap Electrodes

Electrochemical deposition methods, combined with standard lithography techniques, provide a simple, accurate, and more reproducible way for the fabrication of the nanogap electrodes [41-45]. The initial electrodes with a relatively large gap are fabricated by conventional lithography techniques on substrates. The gap is then narrowed down to few nanometers by electrochemically depositing specific atoms from a solution onto the lithographically defined electrodes (Fig 2(b)). This method can conveniently prepare gaps that range from several angstroms to 10 nm. Another advantage is the ability to fabricate asymmetric nanogap electrodes [46, 47]. Chen et al. have recently reported a method for fabricating heterometallic nanogaps, which are made of two different metal nanorods separated by a nanometer-sized gap [48].

3.3 Oblique Angle Shadow Evaporation

The oblique angle shadow evaporation method is often combined with optical and electron-beam lithography to define metal leads [49, 34, 50, 51]. By suspending the mask above the substrate and controlling the direction of the deposition angle with respect to the surface normal, features with sizes smaller than those of the masks can be obtained (Fig. 2(c)). Gap electrode separations smaller than 10 nm can be reproducibly fabricated in this way [51]. Moreover, the gap size could also be adjusted by changing stepwise the tilt angle. Sun and coworkers [50] have achieved the fabrication of large nanojunction arrays with gaps as small as 3 nm and successfully applied them to electrical studies on nanocrystals.

3.4 Electromigration and Electrical Breakdown Methods

Electromigration has also been utilized successfully to fabricate nanogap electrodes for nanodevices [52, 27-29, 53-55]. The flow of a large current density or application of a dc voltage to a thin metal wire predefined by electron-beam lithography causes the electromigration of metal atoms and eventual breakage of the nanowire [54]. This process can yield a stable electrode separation of 1 nm with high efficiency (Fig 2(d)) and is especially advantageous to prepare three-terminal devices and electrode arrays [29]. Recently, Xiang and co-workers [53] have reported the fabrication of sub-5 nm reconnectable nanogaps by a reproducible and automated, feed-back controlled electromigration process. By applying a simple voltage ramp, nanogaps were electrically reconnected many times. Ittah et. al. have [56] reported fabrication of metal quantum point contacts (MQPCs), with dimensions comparable to the de Broglie wavelength of conducting electrons, revealing ballistic transport of electrons and quantized conductance values in the range of $1-4G_0$ ($G_0 = 2e^2/h$), with a contact stability of hours at room temperature . The contacts can be deterministically switched between conductance values, or reform in case they break, using voltage pulses. However the electromigrated nanojunctions have to face two major issues. The first one is the role of temperature in the breaking of a nanowire. The other major problem is the possible metal debris left behind in the gap after fabrication, which can interfere with the introduction and the following studies of transport through metal nanoislands (or other nanosystems of interest) [57].

3.5 Template-based methods

Haztor and co-workers [58] had presented a method of constructing 30-nanometer structures in close proximity with precise spacings that uses the step-by-step application of organic molecules and metal ions as size-controlled resist on predetermined patterns, such as those formed by electron-beam lithography. The organic molecules serve as a ruler for scaling down a larger "parent" structure. After metal deposition and lift-off of the organic multilayer resist, an isolated smaller structure remains on the surface. This approach was first used to form thin parallel wires (15 to 70 nanometers in width and 1 micrometer long) of controlled thickness and spacing. Since then this method has been readily developed further and has been in use for the realization of nanogap electrodes [59-61].

3.6 Large scale nanojunction arrays by optical lithography and wet etching

Fabrication of large arrays of nanojunctions by optical lithography and wet etching of an AlGaAs/GaAs quantum-well (QW) structure was first reported by Krahne et al. [62]. The main advantage in this method is that the thickness of the quantum well and of the deposited metal layer controls the nanogap size with sub-nanometer precision, without the need of expensive e-beam systems [63, 64]. However, this innovative approach to nanoscale electronics had some intrinsic drawbacks: because of the bulk leakage currents through the semiconductor substrate under ambient conditions, such nanojunctions could be only potentially employed at cryogenic temperatures and in the dark. Later Maruccio et al. [65-67] significantly improved this method by means of selective oxidation of the two AlGaAs barriers above and below the QW in order to reduce the leakage current through the semiconductor layers and thus extended the functionality of these nanoscaled devices to room temperature [Fig 2 (e)]. Typical molecular/nanocrystal features, such as Coulomb blockade, Coulomb staircase, rectification, negative differential resistance (NDR) and other magnetoelectrical behaviour were successfully studied on these systems [62, 65]. This method provides a simple, highly reproducible, and economical way of fabricating metallic electrodes separated on the nanoscale and it is especially promising for the construction of large arrays of nanojunctions.

3.7 Other fabrication methods

Recently there has been a hunt for other fabrication methods to produce nanogap devices like focused ion beam (FIB) lithography [33]. Because the minimum diameter of a FIB is relatively small (approx. 5–20 nm) and the proximity effects are much less than for electron beam lithography, a FIB could be a reliable way to prepare nanogap electrodes with the advantages of maskless nanostructuring, good reproducibility and high speed of the nanogap cutting. Similar to the FIB technique, oxidative plasma can also be utilized for elaborate ablation of pre-patterned nanowires to form nanogap electrodes [68, 69]. Other methods include scanning probe lithography (SPL) where a microscopic or nanoscopic stylus is moved mechanically across a surface to form a pattern [70]. Because the sharp tips of a scanning probe microscope with a radius smaller than 10 nm can be neatly and controllably manipulated across the substrate, SPL can be utilized to construct nanogap electrodes [71].

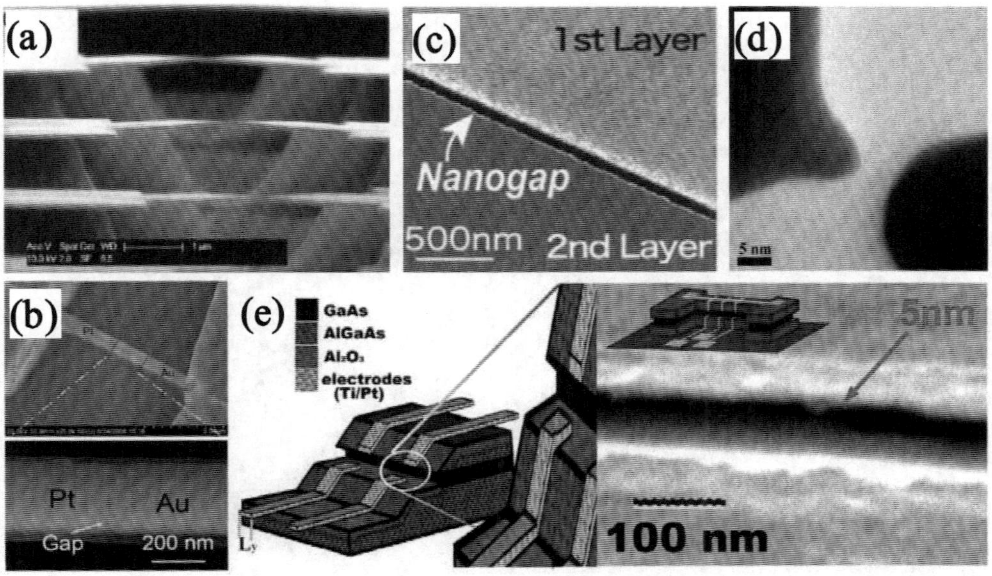

Figure 2. (a) SEM image of a MCBJ. Reproduced with permission from Martin et al., New J. Phys. 10 (2008) 065008 Copyright © 2008 IOP. (b) SEM image of an on-wire lithography 3 nm Au-Pt heterometallic nanogap. Reproduced with permission from Chen et al., Nano. Lett. 9 (2009) 3974 Copyright © 2009 ACS. (c) SEM image of the nanogap between first- and second-layer electrodes. Reproduced with permission from Yoichi et al., Nanotechnology 15 (2004) 1639 Copyright © 2004 IOP. (d) TEM images of a typical electromigrated nanogap on SiN_x membrane. Reproduced with permission from Strachan et al., Nano Lett. 6 (2006) 441 Copyright © 2006 ACS. (e) Schematics and optical micrograph of the mesa nanojunctions. Adapted with permission from Maruccio et al. Small 3 (2007) 1184 Copyright © 2007 John Wiley & Sons.

4. Positioning the nanoparticles

A prerequisite for future applications using nanoparticles as functional entities is control of their arrangement, since they typically require to be located to appropriate substrate locations in order to be connected among themselves and to the outside world. This is still an extremely challenging goal and different approaches are possible to accomplish the controlled positioning.

Molecular, physical, or electrostatic templates can be used for precise positioning of the active nanomaterials. Self-assembled monolayers (SAMs) are ordered assembly of organic molecules that spontaneously form on the surface of metals, metal oxides, and semiconductors. The surface properties of SAMs can be modulated by selecting appropriate tail groups or modifying them with various techniques [72-74]. Then, the substrate surface functionalized with localized patterns of SAMs can serve as template onto which nanoscale or microscale building blocks are selectively attached (Fig 3 (a,b,f)). Physical templates includes holes and ditches that can be fabricated on a substrate surface using lithography and etching/lift-off techniques [75]. The general methods are capillary force driven placement into physical templates, where the capillary force is exploited to place individual nanoscale materials blocks into trenches or holes (Fig. 3c) [75]. Another example of the attaching strategy is dip-pen nanolithography (DPN), which was pioneered by Mirkin and co-workers [76, 77] and uses an atomic force microscope (AFM) tip to transport adsorbed molecules to precise substrate locations with resolution as high as a few tens of nanometers.

Electrostatic interactions between a charged substrate and nanoscale building blocks can be also employed for controlled positioning by letting the building blocks interact with electrostatic templates created using microcontact printing, electron beams, ion beams and scanning probe microscopes [76, 77]. Placement via dielectrophoresis exploits another completely different approach, consisting of motion of uncharged objects in a liquid dielectric medium under a non-uniform electric field [78-81], which can be successfully controlled by a suitable design. Recently [81], a lot of efforts has been made to utilize dielectrophoresis for controlled positioning of nanomaterials for realization of nanoelectronic devices or sensors (Fig. 3 (d,e)). This process is also suitable for placement of the building blocks onto addressable locations on a large scale.

Figure 3. (a) AFM images of nanoscale gaps after nanoparticle self-assembly. Reproduced with permission from Chu et al., J. Am. Chem. Soc., 129 (2007) 2287 Copyright © 2007 ACS (b) SEM image of a nanodevice with immobilized nanoparticles. Reproduced with permission from Kuemmeth et al., Nano Lett., 8 (2008) 4506 Copyright © 2008 ACS. (c) Low-magnification and high-magnification SEM images of single 50-nm Au nanoparticles organized between nanojunctions. Reproduced with permission from Cui et al. Nano Lett., 4 (2004) 1093 Copyright © 2004 ACS. (d, e) Typical DEP assembled gold nanoparticle structures between different electrodes. Fused or melted nanoparticle interconnect and half-melted/unmelted nanoparticle interconnect. Reproduced with permission from Xiong et al., Appl. Phys. Lett. 91 (2007) 063101 Copyright © 2007 AIP. (f) Mesa nanojunctions with immobilized DNA probes after hybridization with gold NPs functionalized with complimentary DNA strands: a number of nanoparticles immobilized on the electrodes and close to the gap are visible. Reproduced with permission from Maruccio et al., Analyst 134 (2009) 2458 Copyright © 2009 The Royal Society of Chemistry.

5. Transport and magnetotransport through metal and magnetic nanoparticles

The focus of this review is the transport behaviour of nanocrystals or quantum dots, which are quasi-zero-dimensional nanostructure systems and whose electronic states are completely quantized. The confinement of carrier motion in these structures is imposed in all three spatial directions, resulting in a discrete spectrum of energy levels as in an atom or molecule. This analogy if further reinforced by the presence of atom-like wavefunction in semiconductor quantum dots and nanocrystals [10-12]. In these systems the introduction of even a single electron is sufficient to dramatically change the transport properties due to the charging energy associated with this extra electron. One of the main consequences is to give rise to Coulomb blockade of transport, where conductance oscillations are found on addition or subtraction of a single electron from a quantum dot [4, 5]. In metal nanoparticles, the resulting transport behaviour can be analyzed purely in terms of classical charging. On the other hand, the study of single-electron tunnelling in semiconductor quantum dots and nanoparticles allows the observation of new phenomena, not observable in their metallic counterparts, which arise from the fact that strong quantum confinement in the low-density limit leads to quantized energy levels whose energy scales can be comparable to the Coulomb charging energy [5-12, 82, 83].

5.1 Basic Principle

The equivalent electrical circuit which reveals the properties of single-electron tunnelling comprises of only one Coulomb island and two leads/electrodes (Figure 1). A third gate electrode, capacitatively coupled to the island, may be added to this double tunnel junction system [84, 5]. Applying a bias voltage between the electrodes, may cause either sequential transfer of electrons into and out of the central island or no charge transport at all. The result depends on the drain-source voltage V_{ds} (or V) as well as on the voltage V_g applied to the gate electrode. The total energy change of the system ΔE while one electron is tunnelling in one of the junctions consists of the charging energy of the island itself as well as of the work done by the voltage source i.e. $\Delta E = e^2/2C \pm eV/2$. This means that for $-e/C \leq V \leq e/C$ the change of energy is positive. Hence, electron tunnelling would only increase the energy of the system, and this transition does not occur if the system cannot take some energy from its environment, i.e., when the temperature is assumed to be low enough. Therefore there is a Coulomb blockade state in the single-electron transistor when the voltage V is within the interval given previously and V_g is zero. Outside this range, the device conducts current by means of sequential tunnelling of electrons. If a finite gate voltage V_g is applied, an additional term has to be included in the energy, because of additional polarization of the island capacitively/electrostatically induced by the gate. The result of this consideration is illustrated in Figure 4(a), where the V vs V_g diagram is shown with its typical CB diamonds. The behaviour of the transistor outside the Coulomb blockade region also shows the single-electron peculiarities, especially for the case of a highly asymmetrical junction, where $R_1 >> R_2$, and $C_1 >> C_2$. In Figure 4(b), one can see the I(V) characteristics with a step-like structure fading with decrease of V. This so-called Coulomb ''staircase'' results from the fact that an increase in the total energy increases the number of channels for tunnelling in a step-like manner, allowing an increasingly larger number of electrons to be present on the island. Another demonstration of charging effects in the SET is the offset of the linear asymptotes by e/C. The dependence of the transistor current on the gate voltage opens up the opportunity to fabricate a sensitive device which measures either directly an electric charge on the island or the charge induced on the island by the charges collected to the gate, i.e. a highly sensitive electrometer. The sensitivity of such an electrometer, which has already been reached in practice, is of the order of parts of electronic charge, exceeding by far the charge sensitivity of conventional devices [85-87].

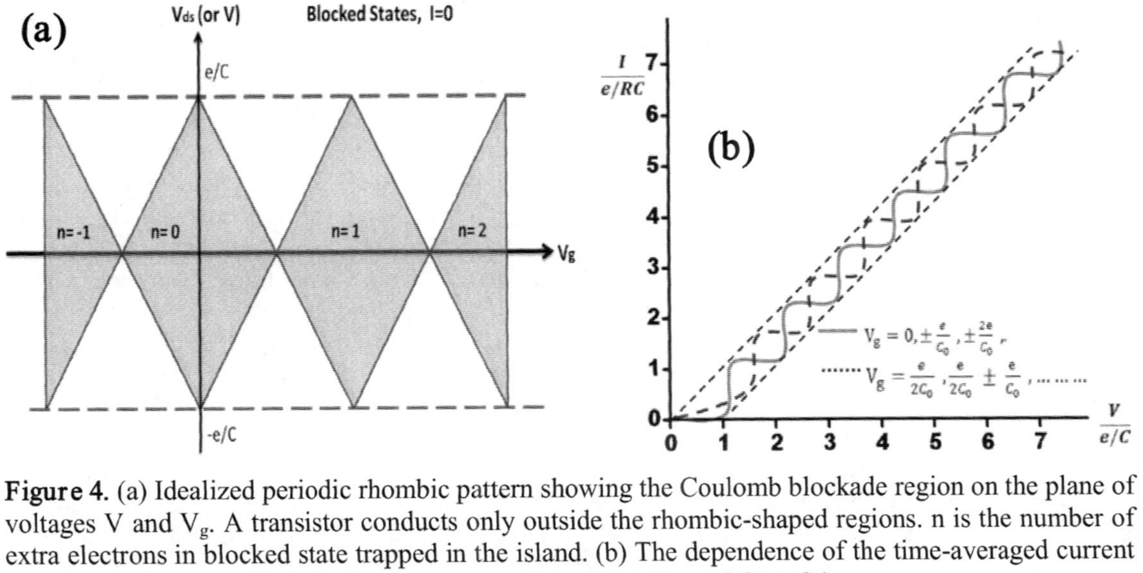

Figure 4. (a) Idealized periodic rhombic pattern showing the Coulomb blockade region on the plane of voltages V and V_g. A transistor conducts only outside the rhombic-shaped regions. n is the number of extra electrons in blocked state trapped in the island. (b) The dependence of the time-averaged current I vs V for an asymmetrical single-electron transistor ($R_1 >> R_2$, and $C_1 >> C_2$).

5.2 Experiments on Single-Electron Tunnelling in Nanoparticles

The theoretical background of SET has found its first experimental manifestation in lithographically fabricated capacitors, having typically capacitances in the pF range [17, 25, 37]. Because of their extremely low charging energies, they needed to be cooled down to very low temperatures for single-electron operation. Furthermore, such a capacitor is typically driven at an operating voltage of 10–100 mV, which leads to the storage of a few tens of thousands of electrons per charging. The utilization of SET events for applications up to the range of room temperature leads to the necessity of decreasing the junction capacitance, at least by two orders of magnitude. As already discussed in above sections, this can be realized by the use of ligand-colloidal metal nanoparticles in a size range of few nanometers. These nanoparticles are stabilized by organic molecules, which surround the metal cores and play the role of an insulating layer in contacts with neighbouring clusters as well as with conducting objects.

5.2.1 Au nanoparticles

Most of the studies in existing literature on SET incorporating single nanoparticles or island involve gold nanocrystals [80, 88-98, 23, 99-103]. Ralph [104] and co-workers observed well-defined electronic states in a Au nanoparticle using tunnel junctions formed from SAMs at low voltages and mK temperatures. Sato et al. [91] found that a sub-monolayer of gold colloidal particles deposited by using an aminosilane adhesion agent (i.e., 3-2-aminoethylamino propyltrimethoxysilane) transform themselves into chains consisting of a few gold colloidal particles probably linked by alkane chains derived from the employed dithiol molecules. These particle chains were formed on SiO_2 substrates with source, drain, and gate metal electrodes defined by electron beam lithography in order to bridge the source-drain gap with a multi-tunnel junction (with the chain itself). The electron conduction through the chain exhibited a clear Coulomb staircase and periodic conductance oscillations as a function of gate voltage (Fig. 5). These measurement results corresponded closely to simulations based on the orthodox theory [103]. Kim et al. [102] have recently observed Coulomb blockade in the field-emission current of a metallic island between two electrodes freely suspended by thin tunnelling barriers with a third electrode serving as a gating contact to trace the Coulomb staircase of the device. The results were in very good agreement with a theoretical model based on orthodox Coulomb blockade and field emission theory. Similarly, in most of the nanojunctions with metal nanoparticles, Coulomb blockade has been observed with the orthodox theory being the tool to understand the inter-component contribution to the SET [23, 105, 93, 98].

However there are reports [105, 100, 101] where the results could not be explained completely with the orthodox theory. Khosousi et al. [100] fabricated single-electron devices by alkanedithiol assisted self-assembly and found a fivefold discrepancy in single-electron charging energies contrary to predictions of the orthodox model for double tunnel junction devices, determined by Coulomb blockade (CB) voltage thresholds in current-voltage measurements versus those determined by an Arrhenius analysis of conductance in the CB region. The energies do, however, scale with particle sizes, consistent with single-electron charging phenomena. They have suggested that the discrepancy is caused by a multibarrier junction potential that leads to a voltage divider effect. Temperature and voltage dependent conductance measurements performed outside the blockade region are consistent with this picture. Kuemmeth et. al. [105] instead have explained their discrete energy levels of individual gold nanoparticles in a SET with monolayers of organic molecules serving as tunnel barriers by means of random-matrix-theory [106]. In particular, they have shown that the nanoparticles are non-magnetic and have spectra in good agreement with predictions taking into account strong spin-orbit coupling. Also recently, Govor and co-workers [101] have bridged a pair of gold electrodes with a single nanoparticle stabilized with citrate molecules. The device exhibited pronounced linear current–voltage characteristics with switching of the current at a fixed bias voltage ascribed to conformational changes in the citrate molecules induced by charge transfer to the molecules.

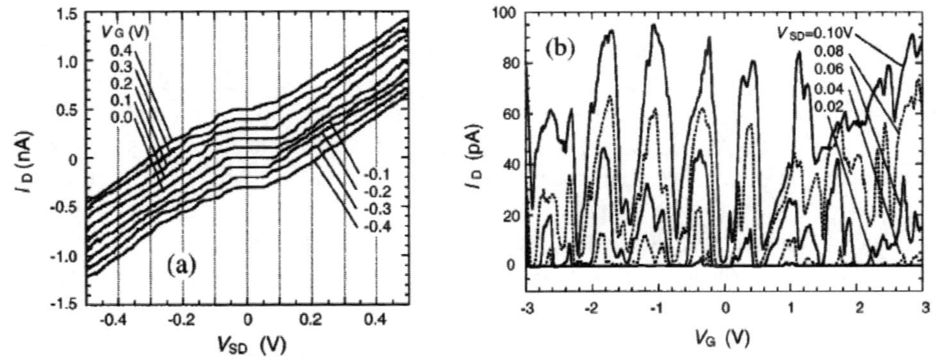

Figure 5. (a) Drain current I_D vs source-drain voltage V_{SD} characteristics at various gate voltages V_G and (b) I_D-V_G characteristics at various source-drain voltages V_{SD} of a three-dot gold colloidal particle transistor measured at 4.2 K . Reproduced with permission from Sato et al., J. Appl. Phys. 82 (1997) 696 Copyright © 1997 AIP.

5.2.2 Other metallic, magnetic and semiconducting nanoparticles

Although the majority of the existing literature deals with studies of Coulomb blockade and SET operation with gold nanoparticles, there is also a considerable attention to other systems like other metallic nanoparticles [107-110], magnetic nanocrystals [111-114] and semiconducting quantum dots [115-121]. Ralph et al. [107] investigated the spectrum of discrete electronic states in single, nm-scale Al particles incorporated into a new kind of tunnelling transistor. The addition of the gate allowed (i) measurements for different numbers of electrons in the *same* particle, (ii) a greatly improved resolution and qualitatively new results for spectra within superconducting particles, and (iii) detailed studies of the gate-voltage dependence of the resonance level widths, which have directly demonstrated the effects of non-equilibrium excitations. Wolf and co-workers [115] have instead very recently described an approach to fabricate single-electron devices consisting of a silicon quantum dot (QD) between metallic leads. The devices showed Coulomb blockade corresponding to a charging energy of 19.4 meV and could be switched from a non-conducting to a conducting state giving rise to Coulomb diamonds. The behaviour was found to be well reproduced by a numerical orthodox theory calculation similar to metallic systems. Recently, Hai and co-workers [116] have reported that spin relaxation can be strongly suppressed for extra electrons confined in a ferromagnetic-metal nanoparticle, resulting in a very long spin-relaxation time of up to 10 μs (Fig. 6). They have suggested that such a long spin-relaxation time could be useful in nanoscale metallic spintronic devices, such as spin memory and single electron spin transistors.

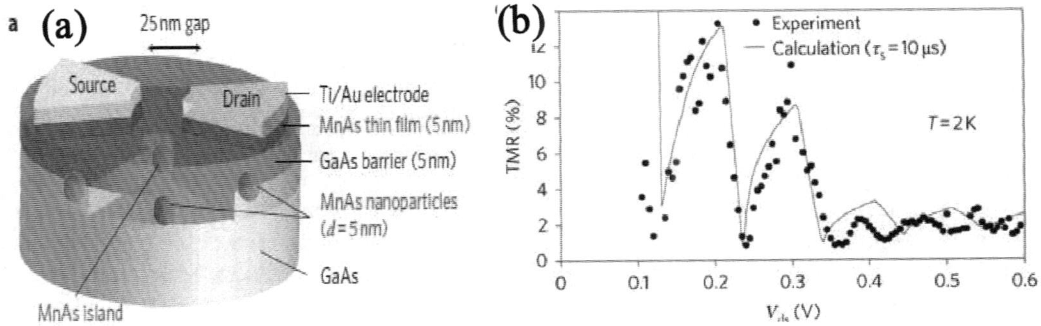

Figure 6. (a) Schematic of a device structure with a MnAs nanoparticle as a ferromagnetic island. (b) TMR = $(R_{H=0} - R_{H=10\,kG})/R_{H=10\,kG}$ as a function of V_{ds}. TMR oscillation was observed up to 0.6 V. Reproduced with permission from Hai et al., Nat. Nano, 5 (2010) 593 Copyright © 2010 Macmillan Publishers Limited.

The use of magnetic nanoparticles or ferromagnetic leads with metal nanoparticles is stimulating a new field of research, often referred as nanospintronics, since the SET characteristics exhibits even more interesting and potential phenomena with magnetic interactions due to the interplay between spin polarized and single electron tunnelling [122, 123]. Magnetotransport through NPs has been investigated both in vertical structures within granular systems prepared by sputtering or physical evaporation methods [124, 125] and within superlattices made of self-assembled colloidal NCs [126-129]. In 2000, Black et al. reported spin-dependent electron transport in periodic arrays of cobalt nanocrystals with 10% magnetoresistance [130]. An enhanced spin accumulation and magnetoresistance oscillations with periodical sign changes as a function of bias voltage were instead observed in magnetic nanoparticles by Yakushiji et al. [125] as shown in Fig 7 (a-b). Recently Bernard-Mantel and co-workers [131] have found anisotropic magneto-Coulomb effects and magnetic single-electron-transistor action in a single nanoparticle bridging magnetic electrodes. They have demonstrated that a significant magneto-Coulomb effect can be induced by conventional magnetic electrodes and mimic spin-valve magnetoresistance (Fig. 7c-d). They also suggested that the magnetic electrode could act as the gate electrode leading to a ferromagnetic single-electron transistor with only two terminals. On the other hand, Birk et. al.[132] demonstrated a spin-accumulation mechanism of magnetoresistance in an Al nanoparticle with a single ferromagnetic contact. The angle between the accumulated magnetic moment and the magnetization changes with the applied magnetic field, causing magnetoresistance. They concluded that this mechanism needs to be considered when interpreting results in spintronics involving discrete energy levels.

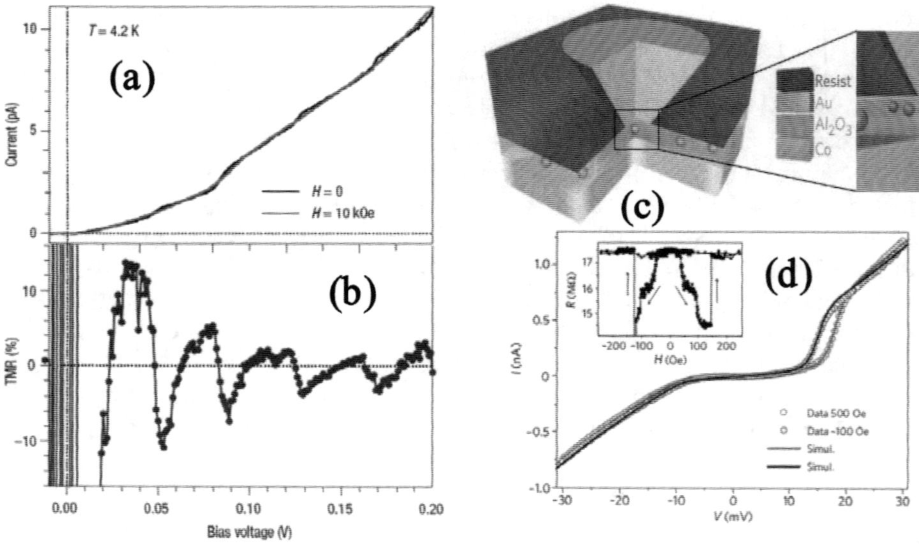

Figure 7. (a) Current–voltage curves and (b) V dependence of the tunnel magnetoresistance, showing oscillation with alternate sign change with applied bias voltage. Reproduced with permission from Yakushiji et al., Nat. Mat., 4 (2005) 57 Copyright © 2005 Macmillan Publishers Limited. (c) Schematic representation of a structure containing a 2D assembly of nanoparticles embedded in a tunnel barrier and connected to top/bottom Co electrodes. (d) Low-bias I–V characteristics at T=1.5 K for the high-resistance state at 500 Oe (red), the low-resistance state at −100 Oe (black) and the corresponding simulations. The inset represents the resistance versus magnetic field measured at a voltage of 14mV. Reproduced with permission from Bernard-Mantel et al., Nat. Phys. 5 (2009) 920 Copyright © 2009 Macmillan Publishers Limited.

Conclusion

Irrespective of their type, nanoparticles generally result in interesting phenomena because of quantum confinement and unique electronic properties which are the basis for future device applications. However the final outcome of this exploration depends on the environment the nanoparticles are experiencing in the nanodevices e.g. the contacts, the protective layer and the assembly. Development in the synthesis of metal nanoclusters and colloidal nanoparticles has resulted in impressive progress in this direction. Chemical processes are also very useful to scale up for industrial purposes. Self-assembly techniques in particular have been so much improved that nowadays semiconductor as well as metal or magnetic nanoparticles can be arranged in perfectly ordered assemblies of remarkable extensions [133]. Much advancement has been accomplished in understanding charge transport phenomena through these nanostructures. The electrical properties of metal nanoparticles are mainly determined by the Coulomb charging energy, and the phenomena can, in general, be understood in terms of single electron tunnelling on the basis of the orthodox theory. It appears that engineered metal nanoparticles in SETs could be a potential approach to further developments of nanoelectronics beyond CMOS technologies in the future. The use of magnetic nanoparticles or ferromagnetic leads then provides advance opportunities in the emerging field of nanospintronics.

Acknowledgements

This work was financially supported by the SpiDME European project (6th Framework Program, NEST, Project No. 029002) and by the Italian Ministry for Foreign Affairs through the bilateral project "Spintronic devices for mass-scale electronics" between Italy and India.

References

[1] Bratton D, Yang D, Dai J and Ober C K 2006 *Polymers for Advanced Technologies* **17** 94
[2] Li T, Hu W P and Zhu D B 2010 *Adv. Mater.* **22** 286-300
[3] Maruccio G and Bramanti A 2009 *Nanobioelectronics – for Electronics, Biology, and Medicine,* ed A Offenhäusser and R Rinaldi: Springer New York) pp 139-66
[4] Likharev K K 1999 *Proceedings of the IEEE* **87** 606-32
[5] Maruccio G and Wiesendanger R 2010 *Quantum Materials, Lateral Semiconductor Nanostructures, Hybrid Systems and Nanocrystals,* ed D Heitmann: Springer Berlin Heidelberg) pp 183-216
[6] Banin U, Cao Y W, Katz D and Millo O 1999 *Nature* **400** 542-4
[7] Liljeroth P, van Emmichoven P A Z, Hickey S G, Weller H, Grandidier B, Allan G and Vanmaekelbergh D 2005 *Physical Review Letters* **95** 086801
[8] Grandidier B, Niquet Y M, Legrand B, Nys J P, Priester C, Stiévenard D, Gérard J M and Thierry-Mieg V 2000 *Physical Review Letters* **85** 1068
[9] Millo O, Katz D, Cao Y W and Banin U 2001 *Physical Review Letters* **86** 5751-4
[10] Maltezopoulos T, Bolz A, Meyer C, Heyn C, Hansen W, Morgenstern M and Wiesendanger R 2003 *Phys. Rev. Lett.* **91** 196804
[11] Maruccio G, Janson M, Schramm A, Meyer C, Matsui T, Heyn C, Hansen W, Wiesendanger R, Rontani M and Molinari E 2007 *Nano Lett* **7** 2701-6
[12] Maruccio G, Meyer C, Matsui T, Talapin D V, Hickey S G, Weller H and Wiesendanger R 2009 *Small* **5** 808-12
[13] Ohnishi H, Kondo Y and Takayanagi K 1998 *Nature* **395** 780-3
[14] Fallahi P, Bleszynski A C, Westervelt R M, Huang J, Walls J D, Heller E J, Hanson M and Gossard A C 2005 *Nano Lett.* **5** 223-6
[15] Agrait N, Yeyati A L and van Ruitenbeek J M 2003 *Physics Reports-Review Section of Physics Letters* **377** 81-279
[16] Morita T and Lindsay S 2007 *Journal of the American Chemical Society* **129** 7262
[17] Li T, Hu W P and Zhu D B *Advanced Materials* **22** 286-300

[18] Lu W and Lieber C M 2007 *Nature Materials* **6** 841-50
[19] Fulton T A and Dolan G J 1987 *Phys. Rev. Lett.* **59** 109
[20] Devoret M H, Esteve D and Urbina C 1992 *Nature* **360** 547-53
[21] Devoret M and Glattli C 1998 *Physics World* **11** 29-33
[22] L. Feldheim D and D. Keating C 1998 *Chemical Society Reviews* **27** 1-12
[23] Ray V, Subramanian R, Bhadrachalam P, Ma L-C, Kim C-U and Koh S J 2008 *Nat Nano* **3** 603-8
[24] Fischbein M D and Drndic M 2006 *Appl. Phys. Lett.* **88** 063116-3
[25] Fischbein M D and Drndic M 2007 *Nano Lett.* **7** 1329-37
[26] Maruccio G, Visconti P, D'Amico S, Calogiuri P, D'Amone E, Cingolani R and Rinaldi R 2003 *Microelectron. Eng.* **67-8** 838-44
[27] Strachan D R, Smith D E, Johnston D E, Park T H, Therien M J, Bonnell D A and Johnson A T 2005 *Appl. Phys. Lett.* **86** 043109
[28] Esen G and Fuhrer M S 2005 *Appl. Phys. Lett.* **87** 263101
[29] Johnston D E, Strachan D R and Johnson A T C 2007 *Nano Lett.* **7** 2774-7
[30] Muller C J, van Ruitenbeek J M and de Jongh L J 1992 *Physica C: Superconductivity* **191** 485-504
[31] Reed M A, Zhou C, Muller C J, Burgin T P and Tour J M 1997 *Science* **278** 252-4
[32] Morpurgo A F, Marcus C M and Robinson D B 1999 *Appl. Phys. Lett.* **74** 2084-6
[33] Tseng A A 2005 *Small* **1** 924-39
[34] Egger S, Ilie A, Fu Y T, Chongsathien J, Kang D J and Welland M E 2005 *Nano Lett.* **5** 15-20
[35] Notargiacomo A, Foglietti V, Cianci E, Capellini G, Adami M, Faraci P, Evangelisti F and Nicolini C 1999 *Nanotechnology* **10** 458-63
[36] Qin L D, Park S, Huang L and Mirkin C A 2005 *Science* **309** 113-5
[37] Qin L D, Jang J W, Huang L and Mirkin C A 2007 *Small* **3** 86-90
[38] van Ruitenbeek J M, Alvarez A, Pineyro I, Grahmann C, Joyez P, Devoret M H, Esteve D and Urbina C 1996 *Review of Scientific Instruments* **67** 108-11
[39] Champagne A R, Pasupathy A N and Ralph D C 2005 *Nano Lett.* **5** 305-8
[40] Christian A M and et al. 2008 *New Journal of Physics* **10** 065008
[41] Kervennic Y V, Van der Zant H S J, Morpurgo A F, Gurevich L and Kouwenhoven L P 2002 *Appl. Phys. Lett.* **80** 321-3
[42] Kervennic Y-V and Kubatkin S 2009 *Small* **5** 2541-4
[43] Boussaad S and Tao N J 2002 *Appl. Phys. Lett.* **80** 2398-400
[44] Qing Q, Chen F, Li P, Tang W, Wu Z and Liu Z 2005 *Angewandte Chemie International Edition* **44** 7771-5
[45] Jemmy S and et al. 2010 *Journal of Micromechanics and Microengineering* **20** 045016
[46] Kashimura Y, Nakashima H, Furukawa K and Torimitsu K 2003 *Thin Solid Films* **438-439** 317-21
[47] Luis De Los Santos V and et al. 2010 *Nanotechnology* **21** 445304
[48] Chen X, Yeganeh S, Qin L, Li S, Xue C, Braunschweig A B, Schatz G C, Ratner M A and Mirkin C A 2009 *Nano Lett.* **9** 3974-9
[49] Yoichi O and et al. 2004 *Nanotechnology* **15** 1639
[50] Sun L F and et al. 2005 *Nanotechnology* **16** 631
[51] Philipp G, Weimann T, Hinze P, Burghard M and Weis J 1999 *Microelectronic Engineering* **46** 157-60
[52] Strachan D R, Smith D E, Fischbein M D, Johnston D E, Guiton B S, DrndiÄ‡ M, Bonnell D A and Johnson A T 2006 *Nano Lett.* **6** 441-4
[53] Xiang C, Kim J Y and Penner R M 2009 *Nano Lett.* **9** 2133-8
[54] Heersche H B, Lientschnig G, O'Neill K, van der Zant H S J and Zandbergen H W 2007 *Appl. Phys. Lett.* **91** 072107-3
[55] Park H, Lim A K L, Alivisatos A P, Park J and McEuen P L 1999 *Appl. Phys. Lett.* **75** 301-3

[56] Ittah N, Yutsis I and Selzer Y 2008 *Nano Lett.* **8** 3922-7
[57] Heersche H B, denbspGroot Z, Folk J A, Kouwenhoven L P, vannbspdernbspZant H S J, Houck A A, Labaziewicz J and Chuang I L 2006 *Phys. Rev. Lett.* **96** 017205
[58] Hatzor A and Weiss P S 2001 *Science* **291** 1019-20
[59] Anderson M E, Mihok M, Tanaka H, Tan L P, Horn M W, McCarty G S and Weiss P S 2006 *Advanced Materials* **18** 1020-2
[60] Li C B and et al. 2010 *Nanotechnology* **21** 495304
[61] Anderson M E, Srinivasan C, Hohman J N, Carter E M, Horn M W and Weiss P S 2006 *Advanced Materials* **18** 3258-60
[62] Krahne R, Yacoby A, Shtrikman H, Bar-Joseph I, Dadosh T and Sperling J 2002 *Appl. Phys. Lett.* **81** 730-2
[63] Sebastian M L and et al. 2005 *Nanotechnology* **16** 1182
[64] Luber S M, Zhang F, Lingitz S, Hansen A G, Scheliga F, Thorn-Csányi E, Bichler M and Tornow M 2007 *Small* **3** 285-9
[65] Maruccio G, Marzo P, Krahne R, Passaseo A, Cingolani R and Rinaldi R 2007 *Small* **3** 1184-8
[66] Maruccio G, Marzo P, Krahne R, Torre A D, Passaseo A, Cingolani R and Rinaldi R 2007 *Microelectronic Engineering* **84** 1585-8
[67] Maruccio G, Primiceri E, Marzo P, Arima V, Della Torre A, Rinaldi R, Pellegrino T, Krahne R and Cingolani R 2009 *Analyst* **134** 2458-61
[68] Whalley A C, Steigerwald M L, Guo X and Nuckolls C 2007 *Journal of the American Chemical Society* **129** 12590-1
[69] Guo X F, Myers M, Xiao S X, Lefenfeld M, Steiner R, Tulevski G S, Tang J Y, Baumert J, Leibfarth F, Yardley J T, Steigerwald M L, Kim P and Nuckolls C 2006 *Proceedings of the National Academy of Sciences of the United States of America* **103** 11452-6
[70] Zhang H, Chung S-W and Mirkin C A 2002 *Nano Lett.* **3** 43-5
[71] Sung M G, Lee T-Y, Kim B, Kim T H and Hong S 2009 *Langmuir* **26** 1507-11
[72] DiBenedetto S A, Facchetti A, Ratner M A and Marks T J 2009 *Advanced Materials* **21** 1407-33
[73] Brust M and Kiely C J 2002 *Colloids and Surfaces A: Physicochemical and Engineering Aspects* **202** 175-86
[74] Chu C, Na J-S and Parsons G N 2007 *Journal of the American Chemical Society* **129** 2287-96
[75] Cui Y, Björk M T, Liddle J A, Sönnichsen C, Boussert B and Alivisatos A P 2004 *Nano Lett.* **4** 1093-8
[76] Piner R D, Zhu J, Xu F, Hong S and Mirkin C A 1999 *Science* **283** 661-3
[77] Salaita K, Wang Y and Mirkin C A 2007 *Nat Nano* **2** 145-55
[78] Bezryadin A, Dekker C and Schmid G 1997 *Appl. Phys. Lett.* **71** 1273-5
[79] Freer E M, Grachev O, Duan X, Martin S and Stumbo D P 2010 *Nat Nano* **5** 525-30
[80] Saiful I K and et al. 2010 *Nanotechnology* **21** 095204
[81] Xiong X, Busnaina A, Selvarasah S, Somu S, Wei M, Mead J, Chen C-L, Aceros J, Makaram P and Dokmeci M R 2007 *Appl. Phys. Lett.* **91** 063101-3
[82] Ploog K H 1993 *Advanced Materials* **5** 227-
[83] Aleiner I L, Brouwer P W and Glazman L I 2002 *Physics Reports-Review Section of Physics Letters* **358** 309-440
[84] Ploog K H 1993 *Advanced Materials* **5** 227
[85] Zimmerli G, Eiles T M, Kautz R L and Martinis J M 1992 *Appl. Phys. Lett.* **61** 237-9
[86] Zorin A B 1996 *Phys. Rev. Lett.* **76** 4408-11
[87] Schoelkopf R J, Wahlgren P, Kozhevnikov A A, Delsing P and Prober D E 1998 *Science* **280** 1238-42
[88] Kano S, Azuma Y, Kanehara M, Teranishi T and Majima Y 2010 *Applied Physics Express* **3** 3
[89] Coskun U C, Mebrahtu H, Huang P B, Huang J, Sebba D, Biasco A, Makarovski A, Lazarides A, LaBean T H and Finkelstein G 2008 *Appl. Phys. Lett.* **93** 123101-3

[90] Yamaguchi H, Terui T, Noguchi Y, Ueda R, Nasu K, Otomo A and Matsuda K 2010 *Appl. Phys. Lett.* **96** 103117-3
[91] Sato T, Ahmed H, Brown D and Johnson B F G 1997 *Journal of Applied Physics* **82** 696-701
[92] Novembre C, Guerin D, Lmimouni K, Gamrat C and Vuillaume D 2008 *Appl. Phys. Lett.* **92** 103314-3
[93] Negishi R, Hasegawa T, Terabe K, Aono M, Tanaka H, Ogawa T and Ozawa H 2007 *Appl. Phys. Lett.* **90** 223112-3
[94] Azuma Y, Yasutake Y, Kono K, Kanehara M, Teranishi T and Majima Y 2010 *Japanese Journal of Applied Physics* **49** 3
[95] Nishino T and et al. 2010 *Nanotechnology* **21** 225301
[96] Wu C S, Chen C D, Shih S M and Su W F 2002 *Appl. Phys. Lett.* **81** 4595-7
[97] Huang S, Tsutsui G, Sakaue H, Shingubara S and Takahagi T 2000 *Journal of Vacuum Science & Technology B: Microelectronics and Nanometer Structures* **18** 2653-7
[98] Persson S H M, Olofsson L and Gunnarsson L 1999 *Appl. Phys. Lett.* **74** 2546-8
[99] Gao B and et al. 2009 *Nanotechnology* **20** 415207
[100] Zabet-Khosousi A, Suganuma Y, Lopata K, Trudeau P-E, Dhirani A-A and Statt B 2005 *Phys. Rev. Lett.* **94** 096801
[101] Govor L V, Bauer G H, Lüdtke T, Haug R J and Parisi J 2010 *Physics Letters A* **374** 3328-31
[102] Kim C, Kim H S, Qin H and Blick R H 2010 *Nano Lett.* **10** 615-9
[103] Hanna A E and Tinkham M 1991 *Physical Review B* **44** 5919
[104] Petta J R, Salinas D G and Ralph D C 2000 *Appl. Phys. Lett.* **77** 4419-21
[105] Kuemmeth F, Bolotin K I, Shi S-F and Ralph D C 2008 *Nano Lett.* **8** 4506-12
[106] Beenakker C W J 1997 *Rev. Mod. Phys.* **69** 731-808
[107] Ralph D C, Black C T and Tinkham M 1997 *Phys. Rev. Lett.* **78** 4087
[108] Wang B, Wang K, Lu W, Yang J and Hou J G 2004 *Physical Review B* **70** 205411
[109] Wang B, Wang K, Lu W, Wang H, Li Z, Yang J and Hou J G 2003 *Appl. Phys. Lett.* **82** 3767-9
[110] Bezryadin A and Dekker C 1997 *Journal of Vacuum Science & Technology B: Microelectronics and Nanometer Structures* **15** 793-9
[111] Gu, eacute, ron S, Deshmukh M M, Myers E B and Ralph D C 1999 *Phys. Rev. Lett.* **83** 4148
[112] Deshmukh M M, Kleff S, Gu, eacute, ron S, Bonet E, Pasupathy A N, von Delft J and Ralph D C 2001 *Phys. Rev. Lett.* **87** 226801
[113] Aleksandrovic V, Greshnykh D, Randjelovic I, Frol̀ˆmsdorf A, Kornowski A, Roth S V, Klinke C and Weller H 2008 *ACS Nano* **2** 1123-30
[114] Petit C, Cren T, Roditchev D, Sacks W, Klein J and Pileni M P 1999 *Advanced Materials* **11** 1198-202
[115] Wolf C R, Thonke K and Sauer R 2010 *Appl. Phys. Lett.* **96** 142108-3
[116] Hai P N, Ohya S and Tanaka M 2010 *Nat Nano* **5** 593-6
[117] van der Wiel W G, De Franceschi S, Elzerman J M, Fujisawa T, Tarucha S and Kouwenhoven L P 2003 *Rev. Mod. Phys.* **75** 1-22
[118] De Franceschi S, van Dam J A, Bakkers E, Feiner L F, Gurevich L and Kouwenhoven L P 2003 *Appl. Phys. Lett.* **83** 344-6
[119] van Kouwen M P, Reimer M E, Hidma A W, van Weert M H M, Algra R E, Bakkers E, Kouwenhoven L P and Zwiller V 2010 *Nano Lett.* **10** 1817-22
[120] Nagamune Y, Sakaki H, Kouwenhoven L P, Mur L C, Harmans C, Motohisa J and Noge H 1994 *Appl. Phys. Lett.* **64** 2379-81
[121] Waugh F R, Berry M J, Mar D J, Westervelt R M, Campman K L and Gossard A C 1995 *Phys. Rev. Lett.* **75** 705-8
[122] Yakushiji K, Mitani S, Ernult F, Takanashi K and Fujimori H 2007 *Phys. Rep.-Rev. Sec. Phys. Lett.* **451** 1-35
[123] Seneor P, Bernand-Mantel A and Petroff F 2007 *Journal Of Physics-Condensed Matter* **19**

[124] Schelp L F, Fert A, Fettar F, Holody P, Lee S F, Maurice J L, Petroff F and Vaurès A 1997 *Physical Review B* **56** R5747

[125] Yakushiji K, Ernult F, Imamura H, Yamane K, Mitani S, Takanashi K, Takahashi S, Maekawa S and Fujimori H 2005 *Nat. Mater.* **4** 57-61

[126] Tan R P, Carrey J, Desvaux C, Grisolia J, Renaud P, Chaudret B and Respaud M 2007 *Physical Review Letters* **99** 176805

[127] Tan R P, Carrey J, Respaud M, Desvaux C, Renaud P and Chaudret B 2008 *Journal Of Magnetism And Magnetic Materials* **320** L55-L9

[128] Poddar P, Fried T and Markovich G 2002 *Physical Review B* **65**

[129] Lekshmi I C, Buonsanti R, Nobile C, Rinaldi R, Cozzoli P D and Maruccio G 2011 *Acs Nano* **5** 1731-8

[130] Black C T, Murray C B, Sandstrom R L and Sun S H 2000 *Science* **290** 1131-4

[131] Bernand-Mantel A, Seneor P, Bouzehouane K, Fusil S, Deranlot C, Petroff F and Fert A 2009 *Nat Phys* **5** 920-4

[132] Birk F T and Davidovicacute D 2010 *Physical Review B* **81** 241402

[133] Pileni M P 2008 *Journal Of Physics D-Applied Physics* **41**

Journal of Physics: Conference Series **292** (2011) 012003 doi:10.1088/1742-6596/292/1/012003

LSMO - growing opportunities by PLD and applications in spintronics

M Cesaria[1, 3], **A P Caricato**[1], **G Maruccio**[1,2] **and M Martino**[1]

[1] Physics Department, University of Salento, Via Arnesano, 73100, Lecce, Italy
[2] NNL Istituto Nanoscienze-CNR, Via Arnesano, 73100, Lecce, Italy

E-mail: maura.cesaria@le.infn.it

Abstract. Ferromagnetic materials exhibiting at room temperature combination of good conductivity, magnetic and opto-electronic properties are needed for the development of functional spin-devices. Mixed-valence LSMO is an optimal source of fully spin-polarized carriers and shows a rich physics of magnetic phases and transport mechanisms. Many factors, such as growth temperature, oxygen stoichiometry, temperature-dependent oxygen desorption rate, structural matching between the growing film and substrate, film thickness, and defects, influence the LSMO properties. Stabilization of ferromagnetic and conductive behaviours is linked to structural order. Therefore a growth approach allowing congruent deposition of complex materials under controlled, reproducible and tunable conditions is strongly needed. In this respect pulsed laser deposition reveals a well-suited choice. This review aims to give an overview on LSMO thin film properties, deposition and applications, especially in the emerging organic spintronics.

1. Introduction

Spintronics has recently attracted large interest since the use and effective manipulation of the electron spin as a further degree of freedom brings significant advantages and new device functionalities as compared to conventional electronics. Improvements include increased effectiveness in performing mass storage, logic operations and quantum computing as well as non-volatility, low power operation and high switching speeds [1-4]. Recently, research also focused on spin-LED and spin-OLED devices, due for example to the possibility of overcoming the recombination statistic constraints in organic semiconductors through injection of spin-polarized carriers [5-9]. In particular for spin injection and detection in spin-valves as well as efficient spin-OLEDs, it is essential to match crucial requirements such as electrodes exhibiting room temperature ferromagnetism, good opto-electronic properties and work-function alignment with states in the organic layer. In this scenario, optimal magnetic materials for high performance spintronic devices are half-metals as sources of fully spin-polarized carriers [10-13]. In particular, Stoner splitting of spin-up and spin-down subbands can provide half-metallic behaviour to transition metal oxides because of their narrow conduction bands (~1eV) and well known examples are CrO_2, $La_{0.7}Sr_{0.3}MnO_3$ (briefly termed LSMO), the double perovskite Sr_2FeMoO_6 and the ferrimagnet Fe_3O_4 [14]. Among them mixed-valence manganese perovskites were intensively studied since their discovery in the fifties because of their rich physics and unconventional magneto-transport properties, such as ferromagnetism associated to metallic conduction and the occurrence of large negative magnetoresistance [14-17]. For this reason, they

[3] To whom any correspondence should be addressed

attracted a lot of interest for both fundamental studies and technological applications [3, 17-19]. In this context, LSMO is a particularly attractive half-metal manganite and pulsed laser deposition (PLD) [20-23] is a very versatile deposition technique to master film properties. Here, the fundamental and practical issues of the peculiar coupling among charge, lattice and spin which characterizes LSMO are discussed with respect to the interplay among material critical parameters and microstructure. Moreover, selected examples of spintronic devices involving LSMO are reviewed with particular attention to the emerging field of organic spintronics [24], but without the aim to be exhaustive since the field is rapidly growing. In general, the flexible tuning of the LSMO film microstructural features allowed by PLD is crucial in favouring stabilization of the ferromagnetic as well as conductive behaviour.

2. LSMO transport and magnetic properties

Half metallic ferromagnets are ideal materials for spintronics due to their ability to provide nearly 100% spin-polarization which results in effective spin-injection, high magnetoresistance (MR) and an enhancement of all spintronics effects in general [10-13]. As an instance, single spin channel injection is a key chance in developing spin-dependent tunnelling devices [25-28] and to improve the electroluminescence efficiency of OLEDs by tuning recombination statistics [29-32].

Among manganese perovskite oxides $La_{1-x}A_xMnO_3$ (where A= Ca, Sr, Ba) has been widely investigated as an optimal material for spintronic devices thanks to its low carrier density (10^{21}- 10^{22} cm^{-3}), high carrier spin-polarization, highest Curie temperature (T_c = 370K in the bulk) among mixed-valence manganites and colossal magnetoresistivity (CMR) [11, 12, 25, 27, 33, 34]. Figure 1 shows the basic LSMO perovskite crystal structure that, in first approximation, can be viewed as a face centered cubic lattice with a Mn-site at the centre surrounded by six oxygen site anions forming an octahedric structure (oxygen octahedron), and La/Sr cations at the corners of a cube. As matter of fact, the stable unit cell is a distorted face centered cubic lattice structure, due to the occurrence of distorted oxygen octahedron. LSMO structural stability is determined by the charge neutrality criterion and tolerance factor that also depends on the Sr dopant radius. In the LSMO lattice, the rare earth trivalent ions are the larger sized species, while the smaller Mn ions occur in the mixed-valence Mn^{3+}-Mn^{4+}.

In the ideal perovskite structure, the crystal field splits the five *d* orbitals of an isolated Mn ion into a t_{2g} triplet (consisting of the d_{xy}, d_{xz} and d_{yz} orbitals) and an e_g doublet (originating from the d_{x2-y2} and d_{3z2-r2} orbitals) whose degeneracy is further lifted by lattice distortions lowering the crystal symmetry (namely Jahn-Teller distortions, as discussed later). Due to the tetrahedral Mn-O coordination the t_{2g} triplet is energetically lower than the e_g doublet as shown in figure 2(a). The spin resolved band

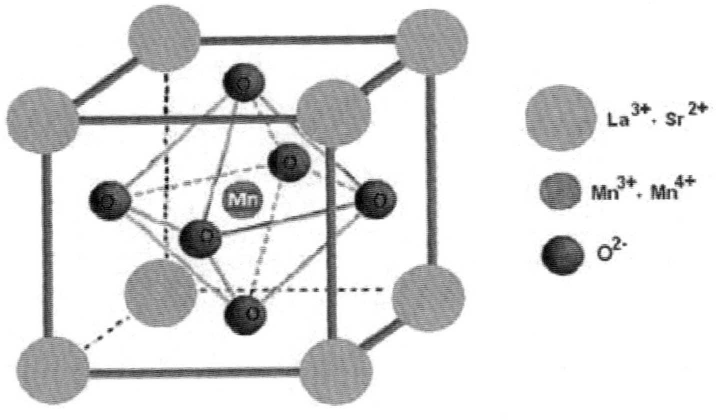

Figure 1: Basic crystal structure of the LSMO perovskite manganite. The oxygen octahedron around the Mn ions in the mixed valence Mn^{3+} to Mn^{4+} is also shown.

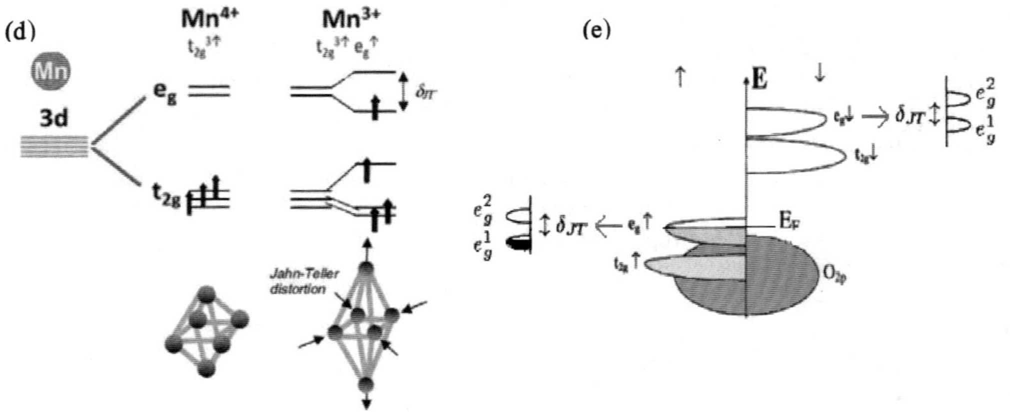

Figure 2: (a) Crystal field splitting of the Mn ion *d*-levels in the perovskite crystal and electronic occupation of the mixed valence Mn ions. The degeneracy of the e_g and t_{2g} levels is lifted by both axial elongation and planar compression as well as axial compression and planar expansion of the oxygen octahedron around each trivalent Mn ion. (b) Sketch of the spin resolved band diagram with lifted degeneracy of the e_g bands.

structure corresponding to this simple level model is sketched [12, 35] in figure 2(b), where the Jahn-Teller distortion-induced lifting of the two-fold degeneracy of the e_g spin-up and spin-down bands is also shown. Hence, in the Mn^{4+} valence states the low-lying t_{2g} states are occupied by three parallel-spin electrons forming a $S= 3/2$ core spin while in the Mn^{3+} valence state (S=2) the additional electron occupies the higher-lying e_g states (with crystal field splitting Δ~1.5 eV between the lowest t_{2g} and the highest e_g level). Basically, substitution of divalent Sr^{2+} for La^{3+} in $LaMnO_3$ causes an equal number of Mn^{3+} to convert to Mn^{4+} and proportionally induces itinerant holes with carrier delocalization in the ferromagnetic phase for a doping concentration of nearly x = 0.3. Doped manganites with x < 0.5 (x > 0.5) have a conduction band more (less) than half filled and are referred to as hole-doped (electron-doped). In hole-doped LSMO the spin-up localized t_{2g} band is fully occupied, whereas the spin-up e_g band is partially occupied and they are completely separated from the empty minority bands by a large Hund's energy (~2.5 eV). The oxygen *2p* states are fully occupied for both spin states and the electronic structure near the Fermi level is determined by the hybridization between the bands associated to the majority spin Mn e_g states and the oxygen *p* states. As an insulating band gap separates the minority-spin-states and the oxygen *2p* band, the LSMO spin-polarized density of states has only majority carriers at the Fermi level. Hence LSMO is a half metal.

As a function of temperature, mixed-valence LSMO shows a metal-insulator (MI) transition as other doped manganites [12, 36, 37]. The metal-insulator transition temperature T_{MI} (that is the peak in resistivity curve versus temperature) and the Curie temperature T_c are both connected to electronic distribution of spin-up and spin-down states and width of the impurity band. Transport regimes and magnetic properties are critically influenced by dopant size and content determining the tolerance factor and average Mn valence, by the relative alignment of the carrier spin to the localized Mn ions and by the structural order related to the growth method and conditions. The two critical temperatures T_{MI} and T_c can coincide, although this is not a universal property of all of the manganites [38].

A qualitative explanation for the metallic conduction below T_c and insulating behaviour above T_c is provided by the "double exchange" (DE) mechanism [39-41] consisting of the ferromagnetic (FM) interaction (through conduction electrons) between the localized 3d magnetic moments with strong on-site Hund's coupling. Accordingly, the relative alignment of adjacent localized spins on the manganese atoms rules the dynamics of itinerant carriers. Charge transport is mediated by the hopping

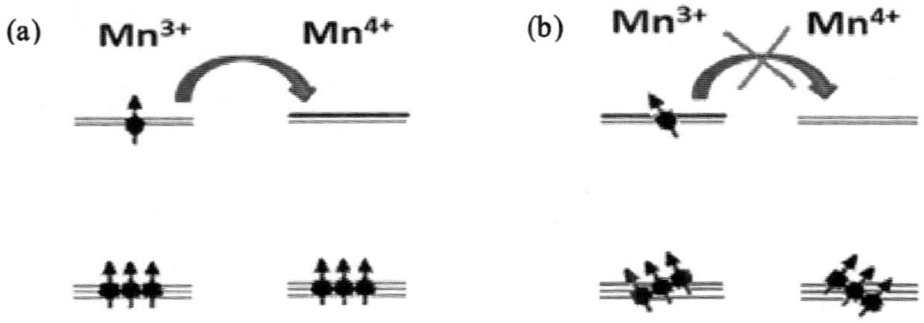

Figure 3: According to the DE model, the electronic transfer probability is (a) maximum when the spin of the Mn^{3+} and Mn^{4+} ions are parallel and (b) increasingly suppressed as the spin misalignment of the Mn^{3+} and Mn^{4+} ions increases.

of electrons between adjacent Mn^{3+} and Mn^{4+} e_g states along the Mn^{3+}-O^{2-}- Mn^{4+} chains. However, the hopping probability is largest when the manganese spins are parallel (figure 3(a)) to each other and smaller otherwise (figure 3(b)). As a result, carriers move easily through the ferromagnetically ordered Mn core spins in the ferromagnetic phase, while in the paramagnetic phase above T_c they are localized due to randomly aligned Mn spins. Consequently, any mechanism capable of suppressing/increasing the spin disorder, such as a magnetic external field or thermal fluctuations, would increase/decrease carrier mobility and lead to a change in the electrical resistivity (e.g. a MR effect). However, this simple spin-dependent electron hopping mechanism fails to explain other experimental observations like the high-temperature insulator-phase [42-44] or the slight difference between the critical temperatures T_{MI} and T_c.

The DE and the localization of itinerant carriers by Jahn-Teller (JT) distortions [45] of the oxygen octahedron surrounding each manganese atom can explain conductivity below T_c and high resistivity around the T_{MI} as well as the observed CMR. Within this model, elongation or compression of the oxygen octahedron occurs due to the spontaneous tendency of trivalent Mn^{3+} to distort its octahedral environment to lift orbital degeneracy. Indeed, the Mn^{3+} degenerate electronic bands e_g make the ideal oxygen octahedron geometrically unstable and cause structural distortions lowering the crystal symmetry. The consequence is a splitting of the e_g and t_{2g} degenerate orbital levels as shown in figure 2(a). In particular, JT distortions lift the two-fold degeneracy of the e_g bands, leading to a doublet of e_g bands which are energetically separated for each spin-channel (figure 2(b)). While the energy of Mn^{4+} remains unchanged, the reduced energy of Mn^{3+} localizes electrons in the paramagnetic phase leading to an insulating behaviour, whereas the increased carrier band width in the ferromagnetic state quenches the JT localization effect inducing metallic conduction.

The JT effect leads to (JT) lattice polarons through a charge-lattice interaction which traps electrons. The concept of polaron is more general since different mechanisms can lead to carrier trapping while the strength of the charge-lattice coupling determines the polaron nature. To explain charge transport and the mid-infrared excitation in the optical conductivity, large polarons (LP) and small polarons (SP) must be invoked [35, 46-48]. Specifically, LP transport is ruled by a self-trapped carrier excited from a trapped ground state to the continuum of unbound states. Conversely, if electron-lattice coupling is sufficiently strong, a SP is excited from its well localized ground state to another well-localized state located on an adjacent site. In both cases, a characteristic energy is needed to either free the self-trapped LP or induce the inter-site SP transition [46, 47]. For high temperatures, the dc-conductivity law for SP hopping is thermally activated over the potential barrier between adjacent sites. In disordered materials, where the carriers are localized by random potential

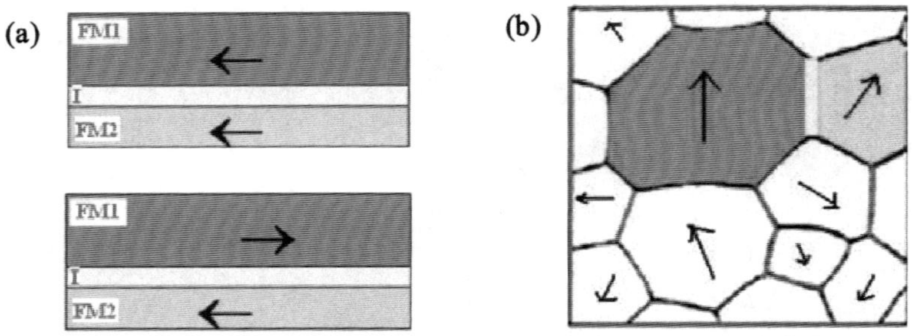

Figure 4: (a) Schematic representation of a spin-valve system: that is a trilayer structure composed of an insulating non-magnetic layer sandwiched between magnetic materials. (b) Magnetic domain structure of a polycrystalline magnet. It is emphasized the generic two-grain system adopted as counterpart of a spin-valve in the SPT and SDS models.

fluctuations, the SP hopping can occur between the nearest sites with large energy differences or between sites beyond the nearest neighbours with small energy difference. In this respect, in polycrystalline manganites the preferentially conductivity regime is the variable range hopping (VHR) corresponding to SP hopping between farther sites with a smaller potential difference and lying within a certain range of energies [48]. As a result, above T_c, an Arrhenius law describes the temperature dependence of the resistivity which exhibits thus a thermally-activated behavior [49-51].

Concerning MR, for single crystals and epitaxial layers MR values approaching to zero at the lowest temperatures and CMR only under application of strong magnetic fields were reported [11, 52, 53]. On the contrary, large low-field MR (LFMR) effects have been reported in polycrystalline bulk and thin films at low temperature [54-58]. These different behaviours observed in the ferromagnetic regime, depending on the microstructure of the material, cannot be explained by the DE mechanism too. The enhanced LFMR response in polycrystalline samples can be ascribed to spin-polarized tunnelling (SPT) between grains [25, 34, 55-57] or spin-dependent scattering (SDS) across interfaces or grain boundaries (GBs) [58-60].

Basically, whereas below T_c the conduction electrons (almost completely polarized inside a magnetic domain) are easily transferred through the Mn–O-Mn chains, spin-disorder dependent scattering at the GBs induces a high zero-field resistivity. Indeed, the magnetization configurations in the polycrystalline materials are differently distributed depending on the grain-magnetic domain structures and magnetically disordered regions occur near the GBs. Such distribution of canted spins acts as insulating and spin-scattering zone. Therefore, electrons can hop from a magnetic domain to another through an intermediate thin insulating layer by spin-dependent tunnelling with a probability essentially depending on the spin states in the neighbouring grains on each barrier side as well as the spin order of the inter-grain zones. This process is similar to transport in a magnetic tunnel junction (spin-valve) as sketched in figure 4, with parallel and antiparallel magnetizations resulting in low resistance and high resistance, respectively. This spin-valve effect allows one to explain the LFMR based on electron flow across GBs and/or inter-grain amorphous phases in polycrystalline manganites (figure 4(b)). A moderate external magnetic field can suppress the spin fluctuations and align grain magnetizations randomly oriented at zero-field. As a result, the resistivity drops and a negative MR is observed, which depends critically on the spin states in the neighbouring magnetic regions and field-induced rotation of the magnetic domains as well as GBs features, occurring of magnetic inter-grain zones and extension of canted spin regions. As a result, the magnetic domain structure of polycrystalline specimens has been extensively studied using different microscopic tools [61-62]. Whereas polycrystalline films have a significantly higher resistivity than their epitaxial counterpart, LFMR in the perovskite manganites $La_{1-x}A_xMnO_3$ with A = Sr, Ca, Ba, and Pb has been emphasized from the point of view of potential applications in digital recording and sensing industries [11, 53, 63].

3. LSMO thin films by PLD and influence of growth conditions

LSMO intriguing properties are strongly related to the lattice-charge-spin coupling, which is very sensitive to phase/structural transformations, oxygen stoichiometry, growth temperature and lattice distortions induced by doping.

Pulsed laser deposition is an ablation technique based on non-equilibrium laser-matter interaction that experienced a rapid spread since 1987 when it revealed very successfully to deposit thin films of high critical temperature superconductors [64]. A crucial advantage of PLD consists of its ability to "transfer" the stoichiometry of complex multi-elemental targets on different substrates. Figure 5 depicts a typical PLD experimental set-up. A short and high-energy laser pulse is focused on a highly absorbing target material. The resulting intense localized heating induces an increase of the local temperature and an instantaneous vaporization/ejection of all the target elements with ionization and formation of a plasma plume. The underlying non equilibrium laser-matter interaction is a process quite different from thermal evaporation where composition depends on the vapour pressures of elements in the target material. In PLD approach, after ablation, plasma species are transferred to the substrate by strongly forward-directed expansion in a controlled background atmosphere [20, 21]. In this respect, ablation can take place either in vacuum or in an inert or reactive atmosphere (reactive PLD) and ambient gas parameters can be tuned to determine the interaction processes and dynamics of the ablated species. For this reason, PLD is the most versatile method for tuning both rate and kinetic energy distribution of the deposition flux [22]. Since growth is driven by periodic bursts of energetic species deposited on a substrate, ad-atoms interact with the growing surface differently from the case of thermal deposition [23]. This allows effective diffusivity regime even at low substrate temperatures, high adhesion to the substrate, epitaxial growth at lower temperature compared to other deposition methods, transition among growth modes, and promotion of the nucleation rate due to the supersaturation and tunable repetition rate of the deposition flux [22, 23]. All these factors are important for the optimal grow of complex materials which, additionally, are sensitive to structural and phase transformations as well as to improve the structural quality and promote very smooth interfaces, which is a basic prerequisite to design functional spin-tunnelling devices.

For LSMO target ablation, basically different laser sources were employed, however, most of them belong to the class of the excimer lasers and the most used are frequency-tripled Nd:YAG laser, KrF (248 nm) or ArF excimer lasers (193nm). Optimal fluence and pulse repetition rate depend on the perovskite oxide to be grown, dopant, substrate nature and deposition temperature. Fluence values

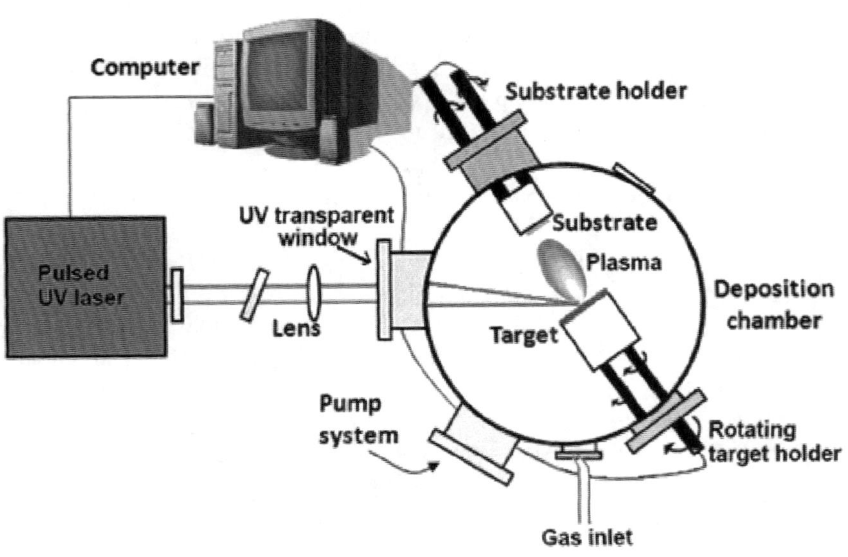

Figure 5: Schematic representation of a standard PLD experimental set-up.

around 4J/cm^2 and 1.5 J/cm^2 are reported in the case of ArF and KrF lasers, respectively, while ablation with Nd:YAG laser is typically performed at 8.5 J/cm^2 for room temperature deposition and at 3 J/cm^2 or lower at higher temperatures. Usual repetition rates range between 5-10 Hz. Reducing the deposition flux frequency favours collisions between oxygen species and ad-atoms migrating on the growing film in between subsequent laser pulses. High deposition temperatures (higher than 600°C) and post-deposition annealing and/or cooling procedures down to room temperature under controlled oxygen pressures are generally mandatory to stabilize structural order, according to the LSMO phase diagram, and obtain maximum MR [38, 54, 56, 65-70]. Raising the substrate temperature provides a driving force for a more effective migration of the adsorbed species, increases the collision rate and nucleation probability and promotes the onset of crystalline phases. However the substrate thermal response plays an important role in determining the maximum allowed as well as the effective growth temperature. Additionally, substrate temperature influences the film oxygen content through oxygen desorption rate. Different parametric studies on the influence of the oxygen pressure on the film magneto-electrical properties are reported in literature but the most used values are around 100 mTorr or 200 mTorr.

Since spintronic applications require T_c and T_{MI} above room temperature, the challenge is to increase both critical temperatures by tuning the deposition parameters. For bulk single crystals and high quality epitaxial films T_{MI} is very close to T_c [36]. On the contrary, in polycrystalline LSMO, T_{MI} is much lower and reducing/oxidizing growth atmospheres result in decreased/increased critical temperatures [71, 72]. The $T_{MI} < T_c$ obtained for granular films has been related to the presence of structural defects and GB regions and depends on growth conditions [73-75]. A $T_{MI} > T_c$ has been also observed in the case of crystalline LSMO films grown by metallorganic chemical vapour deposition [76]. Definitively, whereas no general link can be established between the two critical temperatures, their relationship and separation can be tuned by growth conditions promoting structural order. In this respect, in order to investigate the correlation between T_{MI} and T_c, microstructure, strain and lattice order must be taken into account. As an instance, substrate material, growth temperature and film thickness can induce Curie temperature shift by lattice strain and modified growth kinetics. Basically strain is a function of lattice and thermal expansion mismatch between film and substrate and can lead to distortion and/or rotation of the MnO_6 octahedron. There has been particular interest in the electronic transport and CMR effects of thin films owing to the interplay between substrate-induced strain on the overlayer structure and film thickness [77-79]. In this respect, epitaxial strain imposed by the substrate is an additional degree of freedom for tuning functional properties [80-84]. It has been reported that LSMO thin films epitaxially grown on $SrTiO_2$(001) by PLD can exhibit relatively high room temperature magnetization and $T_{MI} > T_c$, while LSMO films grown onto MgO(001) were found to exhibit nearly equal high critical temperatures. In addition sizable critical temperatures were demonstrated by changing the laser pulse repetition rate and the growth temperature [38]. In this respect, it has been obtained that (i) reducing the former parameter (from 3.3 Hz to 1.3 Hz) at constant substrate temperature (660°C) causes the Curie temperature T_c to increase from 305 K to 340 K, and (ii) increasing the substrate temperature up to 730 °C keeping the repetition rate at 1.3 Hz increases T_c to 350 K.

LSMO deposition by PLD was reported on very different and differently oriented substrates, namely (001) and (100) $SrTiO_3$ (STO), (001) and (100) $LaAlO_3$ (LAO), $YAlO_3$(YAO), (001)-oriented $(LaAlO_3)_{0.3}(Sr_2AlTaO_6)_{0.7}$ (LSAT), (110)-oriented $DyScO_3$ (DSO), $GdScO_3$ (GSO), $NdGaO_3$ (NGO) with orientations (100), (010), (110) and (001), (001) $LaSrGaO_4$(LSGO), (100) Y-stabilized ZrO_2 (YSZ), piezoelectric Pb $(Mg_{1/3}Nb_{2/3})_{0.72}Ti_{0.28}O_3$, crystalline substrates with buffer layers, (001) and (100) MgO, (100) GaAs, (100) Si and amorphous quartz [80, 81, 84, 85-91]. For films epitaxially grown over crystalline substrates, deviations from the bulk lattice constant are observed, depending on the substrate lattice constant and orientation [92]. Substrate-induced stress leads to strong structural variations of the LSMO unit cell influencing the magnetic microstructure of LSMO films which is correlated to the easy axis of magnetization. For example, Kwon *et al.* [80] reported an in-plane easy magnetization direction in tensile-strained films grown on STO(001) substrates and an out-of-plane easy axis in compressively strained films grown on LAO(001) substrates. Furthermore, the magnitude of the tetragonal distortion depends on the relative film–substrate crystallographic orientation with differences observed in both mean grain diameter and film roughness [93, 94].

A gradual change of the grain texture with the substrate temperature during deposition was reported for polycrystalline LSMO films grown on YSZ(100) substrates by PLD with substrate temperature ranging from 600 to 850 °C [72]. Another important aspect is the adatom mobility during the growth on different substrates. The observed similarity in film surface morphology independently from the substrate considered means that during ablation, the good superficial migration allowed by PLD favours both optimal adhesion between film and substrate and strains to be accommodated leading to strong relaxation/distortion of the lattice cell. While for large mismatched MgO substrates the biaxial stress relaxes in a layer of few nanometers, on GaAs lattice mismatch and inter-diffusion make necessary an epitaxial buffer layer but substrate temperature during its deposition should not exceed ~550°C due to threatening decomposition of GaAs. As a consequence, the choice of the buffer-layer material is limited and for this reason MgO is generally employed as buffer layer because it does not require very high deposition temperatures to grow epitaxially and it has better diffusion barrier properties compared with other materials like YSZ. Epitaxial LSMO thin films can be grown on lattice-matched substrates as well as on Si by using lattice-matched buffer layers such as STO, YZS. On the other hand, polycrystalline LSMO films on Si substrates can be obtained by allowing the LSMO to grow on an oxide coating. In this case, growth temperatures below 500°C are considered to allow the preliminary formation of the native oxide on the Si surface. Quartz wafer were also employed as substrate with the aim to avoid the potential effect of crystalline substrates on the film orientation by minimizing the stress effects resulting from epitaxial growth on crystalline substrates.

Another very critical parameter in determining LSMO magnetic and electrical properties is oxygen stoichiometry [72, 85, 95-98]. Basically, each lost oxygen leads to the conversion of two Mn^{4+} ions to Mn^{3+} ions, causing irregular octahedral coordination for Mn atoms and weakening of Mn-O-Mn bondings by lattice expansion. As a consequence, oxygen vacancies strongly reduce carrier density, introduce lattice structural disorder that partially localizes carriers, weaken the DE charge-transfer mechanism through Mn-O-Mn oxygen bridges, decrease both critical temperatures T_{MI} and T_C and increase material resistivity. In contrast, oxydizing conditions lead to ferromagnetic-metallic phase extending up to higher temperatures, due to an increased hole concentration induced by cation vacancies and a decreased Mn-O average distance and the increased Mn-O-Mn bond angles which favour DE. Oxygen stoichiometry depends on growth oxygen pressure, substrate temperature and background atmosphere during post-deposition cooling treatments. Oxygen desorbs from doped manganites at elevated growth temperatures, in a reducing atmosphere as well as during cooling procedures in vacuum. According to the defect model by Roosemalen, under some oxygen partial pressure the excess oxygen concentration decreases with increasing annealing temperature, while at definite annealing temperature the excess oxygen concentration increases with increasing oxygen partial pressure [99-101]. Importantly, oxygen vacancies do not involve irreversible structural disorder: oxygen can be introduced into the sample by annealing the film at high temperatures (~ 400-600°C) in an oxygen-rich atmosphere. When control on oxygen content is desired, PLD is a very versatile technique because increasing the oxygen pressure or reducing the laser frequency during deposition are equivalent ways to promote collisions between oxygen species and the growing film [38, 98].

Finally, the substrate temperature is also important in influencing the film microstructure and, consequently, the film magnetic and electric properties. While thin films grown at low temperature show amorphous or mixed amorphous/ nanocrystalline microstructure, the ones growth at temperature higher than 600°C allows improved crystalline order [102]. Generally for thin films raising the substrate temperature involves a gradual evolution from the PM state to the FM one as well as a crossover from the insulating mode to the metal-like one. This is the result of the increased structural order which stabilizes the ferromagnetic and conductive behaviours.

4. LSMO-based devices: from Spintronics to Organic Spintronics

Spintronics emerged with the discovery of giant magnetoresistance (GMR) by Fert and Grünberg [103, 104] who received the Nobel Prize in Physics in 2007. In magnetic tunnel junctions, the nonmagnetic spacer consists of a thin insulating layer and the MR effect is referred to as tunnel magnetoresistance (TMR). In TMR structures, given the spin conservation during tunneling, the tunnel rate is proportional to the product of the corresponding spin subband DOS at the Fermi level [3]. As a consequence, half-metal LSMO is an optimal material for magnetic tunnel junctions [26]. Moreover, LSMO is very attractive since the conductivity mismatch problem for direct spin-polarized injection at the ferromagnetic metal-semiconductor interface does not apply to fully spin-polarized systems. LSMO has also proved to be a successful spin-injector in artificial grain-boundary devices [105].

The discovery of TMR and GMR in metallic spin-valves meant a revolution in magnetic memories. Very recently the introduction of hybrid spintronic devices consisting of organic materials and conventional magnetic electrodes were another promising turning-point. Organic materials are very appealing because weak spin-orbit and hyperfine interactions in molecules make the spin-relaxation and spin coherence times very long [106]. In organic spintronic devices, organic materials are usually used to decouple two ferromagnetic electrodes while mediating or controlling a spin-polarized signal injected and collected by the ferromagnetic contacts. Relevantly, in this new field, LSMO is the most widely used ferromagnetic spin-injection contact, starting from the first proposed organic spintronic device: an OLED structure consisting of planar LSMO electrodes separated by a nearly 100 nm long channel of α-sexithiophene [107]. In this respect, the low conductivity of LSMO allows to minimize the conductivity mismatch with the organic semiconductor.

Concerning spin valves, prototype hybrid structures based on LSMO and organic semiconductors have been demonstrated [28, 108]. The first functional organic spin-valve [28] consisted of LSMO and Co FM electrodes sandwiching a thick (100-200nm) tris(8-hydroxyquinolinato)aluminium (briefly termed Alq_3) nonmagnetic layer. Figure 6(a) shows the sketch of this large scale vertical device and the bias circuit configuration. The arrow indicates the applied external in-plane magnetic field. Figure 6(b) is a scanning electron micrograph of a LSMO/Alq_3/Co/Al functional spin-valve device. Then, the measured MR response is reported in figure 6(c): the relative magnetization directions of the ferromagnetic electrodes are indicated by horizontal arrows and the sweep direction is shown by the vertical arrows. LSMO spin-injection electrode and Co spin-detector electrode are used as soft and hard magnetic layers. Their different coercitive fields allow one to switch the relative magnetization directions of the ferromagnetic electrodes between parallel and antiparallel alignments, upon sweeping the applied magnetic field. Unlike standard spin-valve devices, the device exhibited lower resistance in the anti-parallel configuration of the electrode magnetizations and this inverse spin-valve effect was attributed to the negative spin polarization of the Co d-band (with a smaller density of states at the Fermi level for the majority spin than for the minority spin subband). An up to 40% negative GMR effect was measured at low temperature (11K). For increasing temperature the MR strongly decreased and such finding was ascribed not to a decreased spin-polarization of LSMO electrodes but to a reduced spin diffusion length.

The same electrodes LSMO and Co have also been used with other organic semiconductors [108]. In this respect figure 6(d) shows a MR loop at low magnetic field measured on a LSMO/CVB/Co spin-valve device at T=14K and with bias voltage of nearly 10mV positively applied to the LSMO electrode. CVB molecule, or 4,4-bis-(ethyl-3-carbazovinylene)-1,1-biphenyl, is an emissive oligomer. Again the device electrical resistance in the case of anti-parallel magnetizations is smaller as compared to the parallel configuration. Switching of the device resistance was confirmed to occur in agreement with the coercitive fields of the LSMO and Co electrodes, namely 50Oe and 600Oe, respectively. The MR value at low magnetic field (<1kG), as estimated by the MR response in figure 6(d), was about 11%. Moreover, a monotonically MR decrease with increasing voltage V was reported.

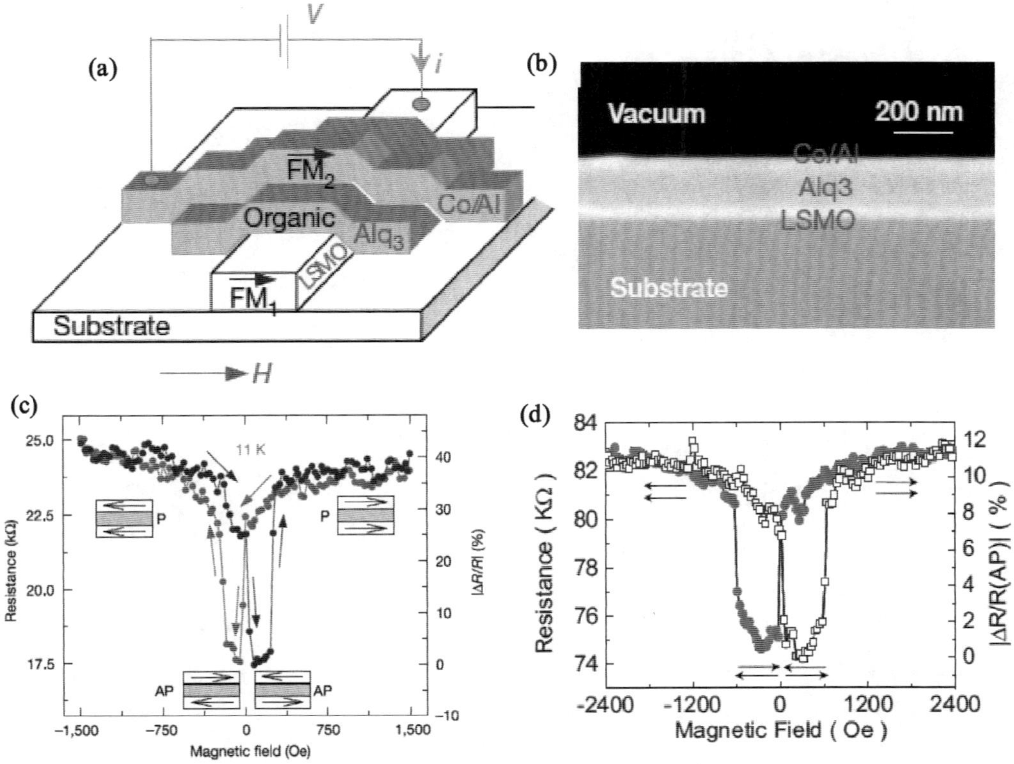

Figure 6: (a), (b) and (c) Sketch, scanning electron micrograph and magnetoresistance response, respectively, of the first functional organic spin-valve demonstrated by Xiong et al. (reproduced with permission from Xiong et al., Nature 427, 821 (2004) Copyright © 2004 Macmillan Publishers Limited). (d) Magnetoresistance measured on a LSMO/CVB/Co spin-valve device at T=14K and with bias voltage of 10mV positively applied to the LSMO electrode (reproduced with permission from Wang et al., Phys. Rev. B 75, 245324 (2007) Copyright © 2007 American Physical Society).

Occurrence of inverse spin-valve effect gave rise to confusion and controversy concerning the sign and mechanisms of MR. Very recently a model invoking the formation of spin-hybridization –induced polarized states in the first molecular layer at the electrode interface has been proposed to explain the observed discrepancies on the MR signs in LSMO/Alq3/Co spin valves [109].

Room temperature MR in LSMO/Alq_3/Al_2O_3/Co devices was demonstrated [110] by improving the quality of both injecting interfaces. Although the observed MR was quite weak, in this case it decreased with temperature in agreement with the surface magnetization of the LSMO and vanished at its Curie temperature [108]. Therefore, the thermal response of LSMO/Co-based spin-valve devices critically depends on the LSMO-interfacial spin-polarization value that vanishes at lower temperature (nearly 220K) compared to the bulk magnetization.

The possibility to inject and transfer spin-polarized currents within carbon nanotubes (CNTs) in a LSMO-CNT-LSMO structure was also demonstrated, along with the transformation of spin information in large electrical signals thanks to the large spin lifetime and mean electron velocity in the channel resulting in large MR [111]. Figure 7(a) and 7(b) show the structure of the junction fabricated by photolithography with LSMO electrodes deposited on lattice matched $NdGaO_3$ substrates and having a 1.5 μm separation and a variable width (1-4 μm, to switch independently their magnetizations). The authors reported tunnel junction-like characteristics and a maximum MR ratio around 60% at 5K (figure 7(d)) with a bias voltage of 25mV, which decreased at larger temperatures until vanishing at 120K (figure 7(e)).

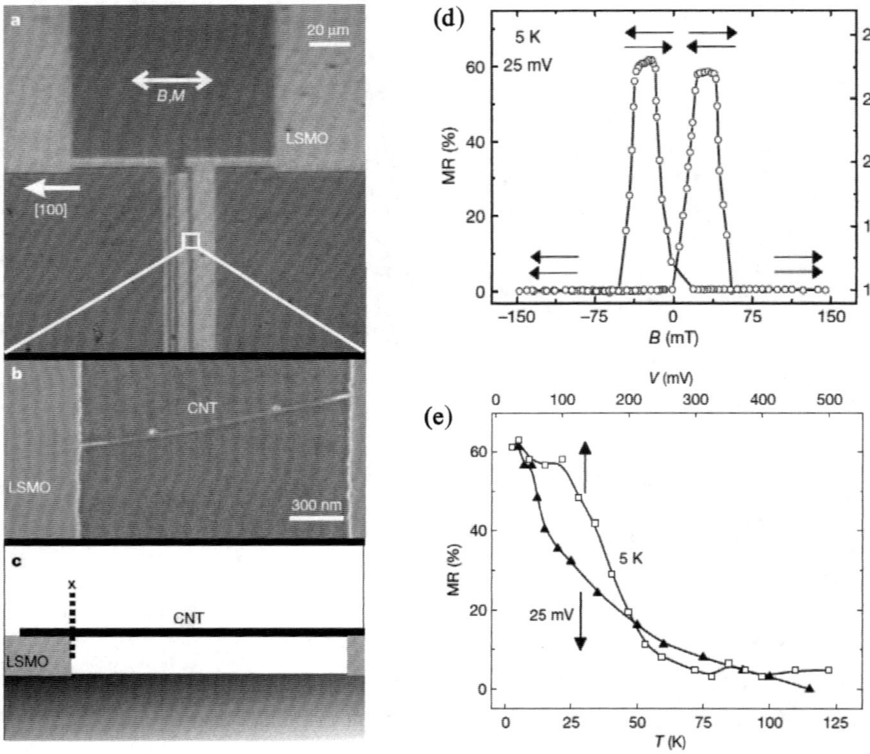

Figure 7: (a) Optical micrograph of four variable-width LSMO electrodes and two of the structure contact pads. Arrows indicate the direction of the in-plane applied magnetic field and the expected direction of magnetization. (b) Scanning electron microscope image of a CNT inserted within the LSMO-CNT-LSMO device. (c) Schematic side view of (b). (d) MR response measured at 5K with a bias voltage of 25mV. (e) MR curves versus temperature and bias voltage. (Figures reproduced with permission from Hueso et al., Nature 445, 410 (2007) Copyright © 2007 Macmillan Publishers Limited).

Very recently, LSMO/C_{60} spin valves have been also implemented. Indeed, devices based on fullerene C_{60} spacer layer are expected to posses thermal stability, namely improved signal/noise ratio of the MR loop measurement [112]. Moreover, while the abundant hydrogen atoms in most organic materials increase the hyperfine coupling, in carbon such coupling occurs only in the rare ^{13}C carbon isotope. Therefore the weak hyperfine interaction of the C_{60} molecule may increase spin lifetimes in organis spin valves based on fullerene.

5. Summary

In summary, the LSMO fundamental magneto-electro-optical properties have been outlined and discussed with respect to their dependence on the deposition parameters and conditions. LSMO has been considered due to its rich physics determined by interplay between charge, lattice and spin interactions. For this reason it is a very interesting material for applications, even in the new field of organic spintronics. Since LSMO structural and magnetic order is very sensitive to the deposition technique, PLD is a very powerful tool to growth high quality films and favour ferromagnetic and conductive phases. In this respect, different independent deposition parameters, namely oxygen pressure, laser fluence and repetition rate, substrate temperature and post deposition treatments, have been discussed concerning their influence on the LSMO properties through the film structure. It has been demonstrated that PLD let the growth of LSMO films with good and very well tailored properties to be used in selected applications.

Acknowledgements. G. Maruccio acknowledges financial support by the SpiDME European project (6th Framework Program, NEST, Project No. 029002) and the Italian Ministry of Foreign Affairs through the high-relevance project for scientific and technological co-operation between Italy and India (Spintronic devices for mass-scale electronics).

References

[1] Wolf S A, Awschalom D D, Buhrman R A, Daughton J M, von Molnar S, Roukes M L, Chtchelkanova A Y and Treger D M 2001 *Science* **294** 1488
[2] Awschalom D D, Loss D and Samarth N (Eds.) 2002 *Semiconductor Spintronics and Quantum Computation* (Berlin: Springer)
[3] Žutić I, Fabian J and Das Sarma S 2004 *Rev. Mod. Phys.* **76** 323
[4] Žutić I 2009 *Nat. Phys.* **5**. 630
[5] Davis A H and Bussmann K 2003 *J. Appl. Phys.* **93** 7358
[6] Shikoh E, Fujiwara A, Ando Y and Miyazaki T 2006 *Jap. J. Appl. Phys.* **45** 6897
[7] Bergenti I, Dediu V, Arisi E, Mertelj T, Murgia M, Riminucci A, Ruani G, Solzi M, Taliani C 2004 *Org. Electr.* **5** 309
[8] Arisi A, Bergenti I, Dediu V, Loi M A, Muccini M, Murgia M, Ruani G, Taliani C and Zamboni R 2003 *J. Appl. Phys* **93** 7682
[9] Dediu V A,. Hueso L E, Bergenti I and Taliani C 2009 *Nat. Mat.* **8** 707
[10] de Groot R A, Mueller F M, van Engen P G and Buschow K H J 1983 *Phys. Re. Lett.* **50** 2024
[11] Jin S, Tiefel T H, Mc Cormack M, Fastnacht R A, Ramesh R and Chen L H 1994 *Science* **264** 413
[12] Park J H, Vescovo E, Kin H J, Kwon C, Ramesh R and Ventakesan T 1998 *Nature* **392** 794
[13] Viret M, Nassar J, Drouet M, Contour J P, Fermon C and Fert A 1999 *J. Magn. Magn. Mat.* **198-199** 5
[14] Ziese M 2002 *Rep. Prog. Phys.* **65** 143
[15] Coey J M D, Viret M and von Molnar S 1999 *Adv. Phys.* **48** 167
[16] Salamon M B and Jaime M 2001 *Rev. Mod. Phys.* **73** 583
[17] Haghiri-Gosnet A-M and Renard J-P 2003 *J. Phys. D: Appl. Phys* **376** R127
[18] Dörr K 2006 *J. Phys. D: Appl. Phys* **39** R125
[19] Bibes M and Barthelemy A 2007 *IEEE Trans. El. Dev.* **54** 1003
[20] Singh R and Narayan J 1990 *Phys. Rev. B* **41** 8843
[21] Amoruso S *et al* 1999 *J. Phys. B: Mol. Opt. Phys.* **32** R131
[22] Eason R (ed.) 2007 *Pulsed Laser Deposition of Thin Films* (Wiley-Interscience)
[23] Aziz M J 2008 *Appl. Phys. A* **93** 579
[24] Naber W J M, Faez S and van der Wiel W G 2007 *J. Phys. D: Appl. Phys* **40** R205
[25] Viret M, Drouet M, Nassar J, Contour J P, Fermon C and Fert A 1997 *Europhys. Lett.* **39** 545
[26] De Teresa J M *et al.* 1999 *Science* **286** 507
[27] Bowen M, Bibes M, Barthelemy A, Contour J-P, Anane A, Lematre Y and Fert A 2003 *Appl. Phys. Lett.* **82** 233
[28] Xiong Z H, Wu D, Vardeny Z V and Shi J 2004 *Nature* **427** 821
[29] Cao Y, Parker ID, Yu G, Zhang C and Heeger A J 1999 *Nature* **397** 414
[30] Friend R ,Gymer RW, Holmes A B, Burroughes J H, Marks R N, Taliani C, Bradley D D C, Dos Santos D A, Bredas J L, Logdlund M and Salaneck W R 1999 *Nature* **397** 121
[31] Bergenti I, Dediu V, Murgia M, Riminucci A, Ruani G and Taliani C 2004, *J. Lumines.* **110** 384
[32] Zhan Y Q, Bergenti I, Hueso L E, Dediu V, de Jong M P and Li Z S 2007 *Phys. Rev. B* **76** 045406-1
[33] Asamitsu A, Moritomo Y, Tomioka Y, Arima T and Tokura Y 1995 *Nature* **373** 407
[34] Bertacco R, Portalupi M, Marcon M, Duo L, Ciccacci F, Bowen M, Contour J P and Barthelemy A 2002 *J. Magn. Magn. Mater.* **242-5** 710
[35] Quijada M, Černe J, Simpson J R, Drew H D, Ahn K H, Millis A J, Shreekala R, Ramesh R,

Rajeswari,M and Venkatesan T 1998 *Phys. Rev. B* **58** 16093 1
[36] Ramirez A P 1997 *J. Phys.: Condens. Matter* **9** 8171
[37] Imada M, Fujimori A and Tokura Y 1998 *Rev. Mod. Phys.* **70** 1039 (and references therein)
[38] Bertacco R, Riva M, Cantoni M,. Signorini L and Ciccacci F 2005 *Appl. Phys. Lett* **86** 252502
[39] Zener C 1951 *Phys. Rev* **81** 403
[40] Anderson P W and Hasegawa H 1955 *Phys. Rev.* **100** 675
[41] de Gennes P G 1960 *Phys. Rev.* **118** 141
[42] Millis A J, Shraiman B I and Mueller R 1996 *Phys. Rev. Lett.* **77** 175
[43] Millis A J, Littlewood P H and Shraiman B I 1995 *Phys. Rev. Lett.* **74** 5144
[44] Millis A J 1998 *Nature* **392** 147
[45] Kugel K I and Khomskii D I 1982 *Usp. Fiz. Nauk* **136** 621
[46] Emin D 1993 *Phys. Rev. B* **48** 13691
[47] Emin D 1994 *Phys. Rev. B* **49** 14290
[48] Viret M, Ranno L and Coey J M D 1997 *Phys. Rev. B* **55** 80678070
[49] Jin K X, Zhao S G and Chen C L 2009 *Physica B* **404** 1515
[50] Jang H P, Shi L, Zhou S M, Zhao J Y, He L F and Jia Y B 2009 *J. Phys.: Condens. Matter.* **21** 046002
[51] Martino M, Cesaria M, Caricato A P, Maruccio G, Cola A (submitted to *Phys. Stat. Sol.*)
[52] Gong C-Q, Canedy C, Xiao G, Sun J Z, Gupta A and Gallagher W J 1995 *Appl. Phys. Lett.* **67** 1783
[53] Hwang H Y, Cheong S-W, Ong N P and Batlogg B 1996 *Phys. Rev. Lett.* **77** 2041
[54] Gupta A, Gong G O, Xiao G, Duncombe P R, Lecoeur P, Trouilloud P, Wang Y Y and Dravid V P 1996 *Phys. Rev. B* **54** R15629
[55] Mathur N D, Burnell G, Isaac S P, Jackson T J, Teo B-S, MacManus-Driscoll J L, Cohen L F, Evetts J E and Blamire M G 1997 *Nature* **387** 266
[56] Liu J -M, Li J, Huang Q, You L P, Wang S J, Ong C K, Wu Z C, Liu Z G and Du Y W 2000 *Appl. Phys. Lett.* **76** 2286
[57] Hwang H Y, Cheong S-W, Ong N P and Batlogg B 1996 *Phys. Rev. Lett.* **77** 2041
[58] Schiffer P, Ramirez A P, Bao W and Cheong S-W 1995 *Phys. Rev. Lett.* **75** 3336
[59] A. Gupta *et al* 1996 *Bull. Am. Phys. Soc.* **41** 555
[60] Raychaudhuri P, Nath T K, Nigam A K, Pinto R 1998 *J. Appl. Phys.* **84** 2048
[61] Liebmann M, Schwarz A, Kaiser U and Wiesendanger R 2005 *Phys. Rev. B* **71** 104431
[62] Taniuchi T, Yasuhara R, Kumigashira H, Kubota M, Okazaki H, Wakita T, Yokoya T, Ono K, Oshima M, Lippmaa M, Kawasaki M and Koinuma H 2007 *Surf. Sci.* **601** 4690
[63] Li X W, Gupta A, Xiao G, Gong G Q 1997 *Appl. Phys. Lett.* **71** 1124
[64] Dijkkamp D, Venkatesan T, Wu X D, Shaheen S A, Jisrawi N, Min-Lee N H, McLean W L and Croft M 1987 *Appl. Phys. Lett.* **51** 619
[65] Mukhopadhyay S, Das I and Banerjee S 2009 *J. Phys. Condens. Matter* **21** 026017
[66] Perna P, Mechin L, Chauvat M P, Ruterana P, Simon C and Scotti di Uccio U 2009 *J. Phys.: Condens. Matter* **21** 306005
[67] Singh U R, Gupya A K, Sheet G, Chandrasekhar V, Jang H W and Eom C B 2008 *Appl. Phys. Lett.* **93** 212503-1
[68] Snyder G F, Hiskes R, DiCarolis S, Beasley M R and Geballe T H 1996 *Phys. Rev. B* **53** 14434
[69] Sadashivan S, Agarwal S, Song T K, Ramesh R, Evans T T, Tuttle B A, Warren W L, Demos D 1998 *J. Appl. Phys.* **83** 2165
[70] Yang F, Kemik N, Biegalski M D, Christen H M, Arenholz E and Takamura Y 2010 *Appl. Phys. Lett* **97** 092503
[71] Wilson M L, Byers J M, Dorsey P C, Horwitz J S, Chrisey D b and Osofsky M S 1997 *J. Appl. Phys.* **81** 4971
[72] Mitchell J F, Argyriou D N, Potter C D, Hinks D G, Jorgensen J D and Bader S D 1996 *Phys. Rev. B* **54** 6172

[73] Mahendrian R, Mahesh R, Raychaudhuri A K and Rao C N R 1996 *Solid State Commun.* **99** 149
[74] Zhang N, Zhang S, Ding W P, Zhong W and Du Y W 1998 *Solid State Commun.* **107** 417
[75] Liu J-M, Huang Q, Li J, You L P, Xu S Y, Ong C k, Liu Z G and Du Y W 2000 *J. Appl. Phys.* **88** 2791
[76] Snyder G Y, Hiskes R, DiCarolis S, Beasley M R and Geballe T H 1996 *Phys. Rev. B* **53** 14434
[77] Wang S H, Wertz E, Hu Y F, Li Q and Schlom D G 2000 *J. Appl. Phys.* **87** 7409
[78] Dho J, Kim Y N, Hwang Y S, Kim J C and Hur N H 2003 *Appl. Phys. Lett.* **82** 1434
[79] Ahn K H, Lookman T and Bishop A R 2004 *Nature* **428** 401
[80] Kwon C *et al.* 1997 *J. Magn. Magn. Mater.* **172** 229
[81] Thiele C, Dörr K, Bilani O, Rödel J and Schultz L 2007 *Phys. Rev. B* **75** 054408
[82] Takamura Y, Chopdekar R V, Arenholz E and Suzuki Y 2008 *Appl. Phys. Lett.* **92** 162504
[83] Chopdekar R V, Arenholz E and Suzuki Y 2009 *Phys. Rev. B* **79** 104417
[84] Tebano A, Orsini, Di Castro D, Medaglia P G and Balestrino G 2010 *Appl. Phys. Lett.* **96** 092505
[85] Garcia V, Bibes M, Barthelemy A, Bowen M, Jacquet E, Contour J P and Fert A 2004 *Phys. Rev. B* **69** 52403
[86] Bergenti I, Dediu V, Arisi E, Cavallini M, Moulin J F, Biscarini F, De Jong M, Dennis C and Gregg J 2005 *Progress in Solid State Chemistry* **XX** 1
[87] Bergenti I, Dediu V, Arisi E, Cavallini M, Arisi E, Riminucci A and Talliani C 2007 *Curr. Appl. Phys.* **7** 47
[88] Bergenti I, Dediu V, Arisi E, Cavallini M, Biscarini F, Taliani C, de Jong M P, Dennis C L, Gregg J F, Solzi M and Natali M 2007 *J. Magn. Magn. Mat.* **312** 453
[89] Calderon S, Escobar-Alarcon V L, Camps E, Muhl S, Rivera M, Bentacourt I, Olaya J and Marifio A 2008 *Microelectron. J.* **39** 1281
[90] Spankova M, Chromik S, Vavra I, Strbik V, Liday J, Vogrincic P, Espinos J P and Lobotka P 2009 *Phys. Status Solid A* **206** 1456
[91] Yang F, Kemik N, Biegalski M D, Christen H M, Arenholz E and Takamura Y 2010 *Appl. Phys Lett.* **97** 092503
[92] Beach G, Yeh N C, Fu C-C, Wei Y T, Vasquez R P, Huynh J, Maurer S M and Beam D A 1997 *J. Appl. Phys.* **81** 5499
[93] Desfeux R, Da Costa A and Prellier W 2002 *Surf. Sci.* **497** 81
[94] Walter T, Dörr K, Müller K-H, Holzapfel B, Eckert D, Wolf M, Schläfer D, Schultz L and Grötzschel R 1999 *Appl. Phys. Lett.* **74** 2218.
[95] Li J, Liu J M, Li H P, Fang H C and Ong C K 1999 *J. Magn. Magn. Mat.* **202** 285
[96] Xu M, Shi Z and Xu Z 2007 *Phys. Stat. Sol.* **204** 1210
[97] Yokota Y, Shimoyama J-I, Ogata T, Horii S and Kishio K 2007 *Solid State Commun.* **142** 429
[98] Signorini L, Riva M, Cantoni M, Bertacco R and Ciccacci F 2006 *Thin Solid Films* **515** 496
[99] Roosmalen J A M and Cordfunke E H P 1995 *J. Solid State Chem.* **110** 106
[100] Roosmalen J A M and Cordfunke E H P 1995 *J. Solid State Chem.* **110** 109
[101] Roosmalen J A M and Cordfunke E H P 1995 *J. Solid State Chem.* **110** 113
[102] Liu J-M, Huang Q, Li J, You L P, Xu S Y, Ong C K, Liu Z G and Du Y W 2000 *J. Appl. Phys.* **88** 2791
[103] Baibich M N, Broto J M, Fert A, Nguyen Van Dau F, Petroff F, Eitenne P, Creuzet G, Friederich A and Chazelas J 1988 *Phys. Rev. Lett.* **61** 2472
[104] Binasc G, Grünberg P, Saurenbach F and Zinn W 1989 *Phys. Rev. B* **39** 4828
[105] Everetts *et al.* 1998 *Philos. Trans. R. Soc. Lond. A* **356** 156
[106] Petta J R, Slater S K and Ralph D C 2004 *Phys. Rev. Lett.* **93** 136601
[107] Dediu V, Murgia M, Matacotta F. C., Taliani C and Barbanera S 2002 *Solid State Commun.* **122** 181
[108] Wang F J, Yang C G, Vardeny Z V and Li X G 2007 *Phys. Rev. B* **75** 245324

[109] Barraud C *et al.* 2010 *Nat. Phys.* **6** 615
[110] Dediu V *et al.* 2008 *Phys. Rev. B* **78** 115203
[111] Hueso L.E, Pruneda J M, Ferrari V, Burnell G, Valdes-Herrera J P, Simons B D, Littlewood P B, Artacho E, Fert A and Mathur N D 2007 *Nature* **445** 410
[112] Wang F J 2009 *PhD thesis*, University of Utah (unpublished)

Journal of Physics: Conference Series **292** (2011) 012004 doi:10.1088/1742-6596/292/1/012004

Electron spin relaxation in organic semiconductors probed through μSR

L Nuccio [1], L Schulz [2], M Willis [1], F L Pratt [3], M Heeney [4], N Stingelin [4], C Bernhard [2], A J Drew [1,2]

[1] Queen Mary University of London, Department of Physics, Mile End Road, London, E1 4NS, UK

[2] Department of Physics & FriMat, University of Fribourg, Ch. du Musée 3, 1700 Fribourg, CH

[3] ISIS Muon Facility, Rutherford Appleton Laboratory, Didcot, OX11 0QX, UK

[4] Centre for Plastic Electronics, Imperial College London, Exhibition Road, London, SW7 2AZ, London, UK

E-mail: `l.nuccio@qmul.ac.uk`

Abstract. Muon spin spectroscopy and in particular the avoided level crossing technique is introduced, with the aim of showing it as a very sensitive local probe for electron spin relaxation in organic semiconductors. Avoided level crossing data on TMS-pentacene at different temperatures are presented, and they are analysed to extract the electron spin relaxation rate, that is shown to increase on increasing the temperature from 0.02 MHz to 0.33 MHz at 3 K and 300 K respectively.

1. Introduction

Organic semiconductors are extremely promising materials for spintronics thanks to their long spin coherence times. However, spin phenomena in these materials are not yet fully understood, partly due to the lack of suitable characterisation techniques, since those tuned for inorganic semiconductors are not always successful for organic materials [1, 2]. Muon spin relaxation (μSR) has been proven to be an applicable probe in this field, as it is sensitive to spin dynamics in organic semiconductors on a molecular lengthscale [3]. In organic semiconductors the implanted positive muons form a hydrogen-like system, called muonium, by capturing an electron. In muonium the muon and electron spins are coupled through hyperfine interaction. As a consequence relaxation on the electron spin can induce changes in the muon spin, which can be detected. Here we show how muon spin relaxation and in particular the avoided level crossing (ALC) technique can be used to probe the electron spin dynamics in organic semiconductors.

2. Muon spin relaxation and avoided level crossings

Muon spin relaxation is an experimental technique using positive muons as microscopic spin probe. Positive muon has a finite lifetime $\tau \sim 2.2$ μs, and it radioactively decays emitting a positron and two neutrinos according to the process $\mu^+ \rightarrow e^+ + \nu_e + \overline{\nu_\mu}$. The positron is emitted preferentially in the direction of the muon's spin at the time of the decay, due to parity violation in weak interaction [4, 5]. As a consequence measuring the time dependence of the emission

direction of the positron provides information on the time evolution of the muon spin.
In μSR experiments 100% spin polarised muons are implanted in the sample. Once implanted in semiconductors the muon gives rise to a series of ionisation and scattering processes, through which its energy decreases from 4MeV to a few keV. Then it takes part in a series of electron capture and loss processes that reduce its energy further. Finally it thermalises to form interstitial positive muon or it captures an electron and form a hydrogen-like species called muonium (Mu) [5]. Thermalisation of muons is fast (<ns) compared to its lifetime.
If an external magnetic field is applied in the direction of the initial muon spin, as it is in so called longitudinal field experiments, the Hamiltonian for muonium can be written as

$$H = -\gamma_\mu \hbar \boldsymbol{I} \cdot \boldsymbol{B} + \gamma_e \hbar \boldsymbol{S} \cdot \boldsymbol{B} + \hbar \boldsymbol{S} \cdot \boldsymbol{A} \cdot \boldsymbol{I} \tag{1}$$

where γ_μ and γ_e are the muon and electron gyromagnetic ratios, B is the applied longitudinal field, I and S are the muons and electrons spins and A is the hyperfine tensor [5]. The energy levels as a function of the applied magnetic field that originates from this Hamiltonian for an isotropic hyperfine interaction are shown in the Breit-Rabi digram in Figure 1(a). At B=0 a

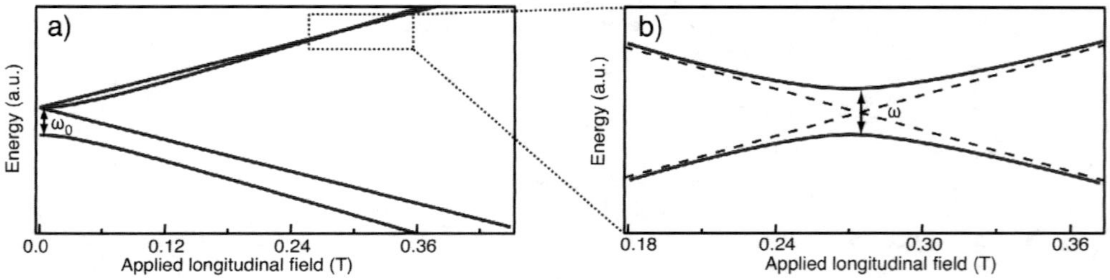

Figure 1. (a) The muonium Breit-Rabi diagram with a hyperfine coupling constant of A=80 MHz (b) Zoom-in region of the energy levels near the ALC.

singlet and a triplet are obtained, the degeneracy of the latter being lifted by the magnetic field.
At high enough fields (Paschen-Back regime, where most of the experiments are performed) the electron and muon spins are decoupled and the state of the system can be described by products of pure Zeeman states for the electron and muon.
At a high field which depends on the value of the hyperfine coupling constant, two of the levels should cross, as can be observed in Figure 1(b) (dashed lines). However, if further interactions are present, as an example with the nuclei, this crossing can be avoided. In fact, for a Hamiltonian $H = H_0 + H^I$ where H^I can be treated as a perturbation, the correction to the energies up to second order is given by

$$E_n - E_n^0 = \langle \chi_n | H^I | \chi_n \rangle + \sum_m \frac{|\langle \chi_n | H^I | \chi_n \rangle|^2}{E_n^0 - E_m^0} \tag{2}$$

where E_n^0 and χ_n are the n-th zero order energy and eigenstate and the denominator becomes tiny when the two unperturbed levels approach [4]. The effect of this correction to the energy levels is to avoid the crossing (See Figure 1(b)). Close to this avoided level crossing the states are no longer pure Zeeman states, but they are instead mixtures of states corresponding to different values of the muon magnetic quantum number. As a consequence a loss of polarisation of the muon can be observed [4, 5]. ALCs can be classified by the change in total quantum number M. If the hyperfine interaction is isotropic, the most common kind of ALC is the one with

ΔM=0, that corresponds to a muon-nuclear spin flip-flop. However in solids the most intense crossing is ΔM=1, corresponding to a muon spin flip and is observed only when anisotropy in the interaction is present. ΔM=2 transitions exist, but they are usually very weak and narrow [3]. A simulated ALC curve is shown in Figure 2 (black line).

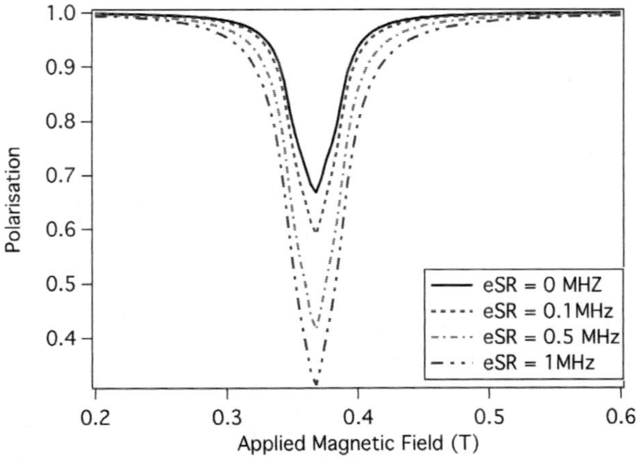

Figure 2. Simulated ALC resonance for different electron spin relaxation rates (eSR)

As can be observed, the polarisation is plotted versus the applied magnetic field. The polarisation exhibits a deepening, representing the loss of polarisation around the ALC field. The shape and position of this ALC curve depends on the hyperfine coupling constants (both isotropic and anisotropic). However, the presence of an electron spin relaxation rate (eSR) also has an effect on the ALC curve. ALCs simulated using the same parameters with an increasing electron spin relaxation rate are shown in Figure 2. It is clear that at low electron spin relaxation rates neither the position nor the shape of these ALCs show substantial changes, but the amplitude enormously increases.

This can be modeled through a density matrix formalism, that is also used to simulate the experimental data and extract the eSR. A density matrix is defined as

$$\rho(t) = \frac{1}{4}[1 + \boldsymbol{p}(t) \cdot \sigma + \boldsymbol{p}_e(t) \cdot \tau + \sum_{j,k} P^{jk}(t)\sigma^j\tau^k] \qquad (3)$$

where $\boldsymbol{p}(t)$, $\boldsymbol{p}_e(t)$ and $P^{jk}(t)$ are the muon, electron and mixed polarisations, respectively. The time evolution of $\rho(t)$ is described by the equation

$$i\hbar\frac{d\rho}{dt} = [H, \rho] \qquad (4)$$

from which the time dependent polarisations can be obtained [6]. The general solution can only be determined numerically. If an electron spin relaxation mechanism is present in the system its effect can be taken into account phenomenologically by introducing extra terms $-\lambda p_e$ and $-\lambda P^{jk}$ in the differential equations for the polarisations obtained from equation 4 [7]:

$$\frac{dp_e}{dt} = \ldots - \lambda p_e, \frac{dP^{jk}}{dt} = \ldots - \lambda P^{jk} \qquad (5)$$

3. Experimental methods

(Trimethyl)silyl-pentacene (TMS-pentacene) was synthesised according to a published procedure [8, 9] and was further purified by repeated recrystallisation from dichloromethano/ethanol. Its structure is shown in Figure 3. The presence of the side groups on the central spine prevents the

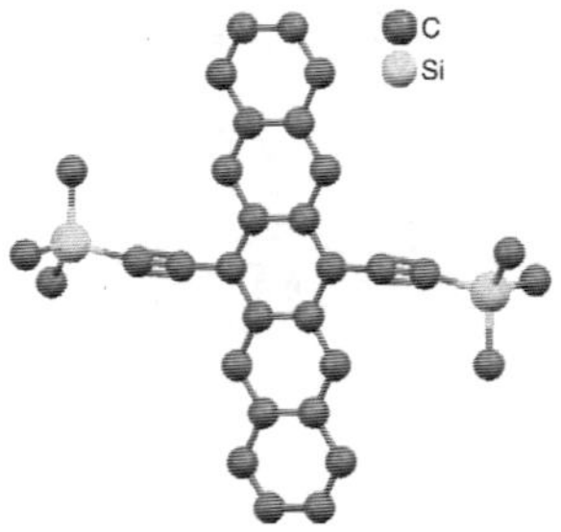

Figure 3. Molecular structure of TMS-pentacene. Hydrogen has been omitted for clarity.

molecule to take the usual herringbone packing of pentacene; as a consequence the molecules have a face-to-face packing that allows better π-orbitals overlap and consequently better charge carrier mobility [8, 9]. The measured sample was polycrystalline powder.
Muon experiments were performed at the HIFI spectrometer at ISIS, Rutherford Appleton Laboratory. About 250 mg of TMS-pentacene was placed in an envelope (17mm $\times$ 17mm) made of 25 μm thick silver foil (99.99 % pure). Another layer of silver foil was placed in front of the sample as a degrader, to assure that the muons stop in the sample.
The software Wimda [10] was used to extract the time-integrated data and plot them as a function of applied field. ALCs were obtained in this way, and they were then modelled using the software Quantum based on the density matrix formalism described in the previous section [11]. The ALC simulations presented in the following are performed considering a muon-electron system. No nuclei are taken into account as their influence was discovered to be negligible. Monte-Carlo simulations with powder averaging and 10000 iterations for each magnetic field was used.

4. Results and discussion

Figure 4 shows the ALCs measured for TMS-pentacene at 3K, 150K and 300K. Data were also acquired at 10K and 25K (data not shown). ALCs are observed around 0.28T, corresponding to a hyperfine coupling of around 70MHz. The position and width of the ALCs do not show a remarkable temperature dependence, while the amplitude considerably increases on increasing the temperature. From the time-differential data and the field dependent relaxation rate (data not shown, see ref [12]) this increased ALC depth is clearly related to a higher relaxation rate of the muon spin.

The measured ALCs look as the superposition of three ALCs very close to each other, the difference arising from slightly different possible bonding angles of the muon in the molecule. As a consequence three contributions were also considered in the modeling. Their relative amplitude is determined from the 3K data, setting the eSR to a very low value. This assumption is justified if the electron spin relaxation is driven by an effect strongly dependent on temperature, as a coupling to vibrations [12]. The result is shown in Figure 4 (black line).The obtained relative amplitudes are 0.17, 0.24 and 0.59. These parameters are used to scale the three contributions at higher temperatures. An eSR different from zero is necessary in the model to account for the higher intensity of the ALC at higher temperatures, whereas there is only a small change in the hyperfine coupling constants. The results of these simulations are shown as solid lines in Figure

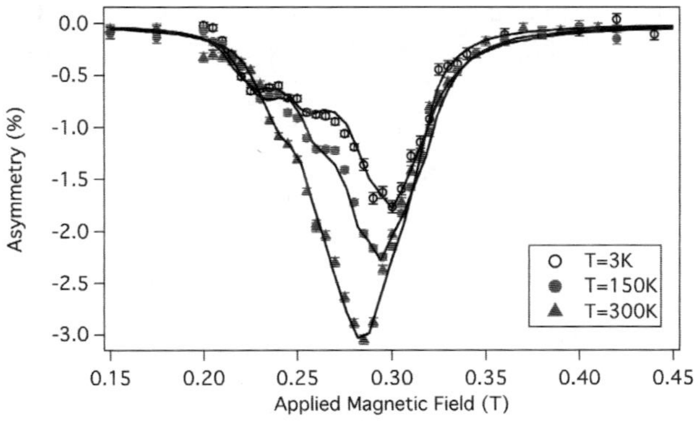

Figure 4. Muon data around the ALC. Black lines are the result of modelling.

Table 1. Simulation parameters for TMS-pentacene.

	3K			**10K**		
	ALC1	ALC2	ALC3	ALC1	ALC2	ALC3
A (MHz)	61.7±0.2	71.2±0.2	81.9±0.2	61.7±0.2	71.2±0.2	81.9±0.2
D1 (MHz)	3.6±0.3	6.0±0.4	5.8±0.4	3.6±0.3	6±0.4	5.0±0.4
D2 (MHz)	3.6±0.5	3.0±0.4	4.0±0.4	3.6±0.5	3.0±0.5	4.0±0.5
eSR (MHz)	0.02±0.02	0.02±0.02	0.02±0.02	0.04±0.02	0.04±0.02	0.04±0.02
	25K			**150K**		
	ALC1	ALC2	ALC3	ALC1	ALC2	ALC3
A (MHz)	62.5±0.2	71.8±0.2	81.6±0.2	63.0±0.2	72.2±0.2	81.0±0.2
D1 (MHz)	3.6±0.3	4.5±0.4	5.0±0.4	3.6±0.3	4.5±0.5	6.0±0.4
D2 (MHz)	3.6±0.5	2.0±0.5	4.0±0.5	3.6±0.5	2.0±0.5	4.0±0.5
eSR (MHz)	0.11±0.02	0.11±0.02	0.11±0.02	0.13±0.02	0.13±0.02	0.15±0.02
	300K					
	ALC1	ALC2	ALC3			
A(MHz)	66.0±0.2	73.0±0.2	79.0±0.2			
D1(MHz)	3.6±0.3	4.5±0.4	6.0±0.4			
D2(MHz)	3.6±0.5	2.0±0.5	4.0±0.5			
eSR(MHz)	0.33±0.02	0.33±0.02	0.33±0.02			

4. Values for hyperfine coupling A, D1 and D2, that are the isotropic, dipolar and non-axial terms, respectively, as defined in Ref. [11], and for eSR as obtained from these simulations are reported in Table 1. The three ALCs tend to merge on increasing the temperature. These small changes in the values of A, D1 and D2, of the order of few percent, are likely to be due to changes in bond length at different temperatures. On the other hand there is a substantial increase in the eSR as a function of temperature, as can be seen in Table 1 and in Figure 5. Figure 5 shows the electron spin relaxation obtained from the simulation as a function of inverse temperature. This behaviour closely resembles the one observed in other organic semiconductors, like rubrene, Gaq_3 and TIPS-pentacene [12]. The electron spin relaxation rate in rubrene and

TIPS-pentacene exhibits an Arrhenius-like temperature dependence, with two characteristic energy scales comparable to the energy of the vibrations in these systems [12]. Data presented here are not detailed enough to extract a full temperature dependence of the eSR, but they are very consistent with [12]. It is also clear that the ALC spectroscopy is a very sensitive technique for the study of the electron spin relaxation in organic semiconductors. Further experiments are needed to get a detailed temperature dependence of eSR in TMS-pentacene.

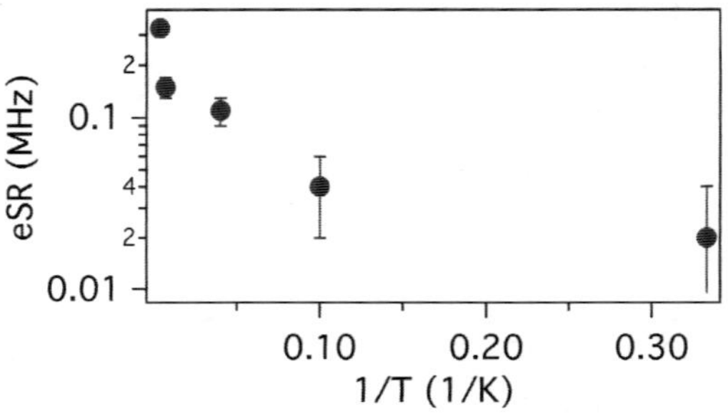

Figure 5. Temperature dependence of the electron spin relaxation rate obtained from the simulations of the ALCs.

5. Conclusions

The present paper reviews the basic physics behind the avoided level crossing technique. It was shown that muons are a very effective probe to measure electron spin relaxation in organic semiconductors thanks to their coupling to the electron via hyperfine interaction. The electron spin relaxation rate can in fact be extracted from ALC measurements. As an example data on TMS-pentacene at three different temperatures are shown and analysed to obtain the electron spin relaxation rate, that is shown to increase on increasing the temperature.

Acknowledgements

We would like to thank E. Roduner, Z. Salman and R. Scheuermann for valued discussions with regards to the analysis and interpretation of our data. AJD acknowledges financial support from the EPSRC (grant EP/G054568/1) and the Leverhulme Trust. CB acknowledges financial support from the SNF (grant 200020-119784 and 200020-129484) and the NCCR program MaNEP.

References

[1] Sanvito S 2007 *J. Mater. Chem.* **17** 4455
[2] Drew A J, Hoppler J, Schulz L, Pratt FL, Desai P, Shakya P, Kreouzis T, Gillin WP, Suter A, Morley NA, Malik VK, Kim KW, Bouyanfif H, Bourqui F, Bernhard C, Scheuermann R, Nieuwenhuys GJ, Prokscha T and Morenzoni E 2009 *Nat. Mat.* **8** 109
[3] Drew A J, Pratt F L, Hoppler J, Schulz L, Malik-Kumar V, Morley N A, Desai P, Shakya P, Kreouzis T, Gillin W P, Kim K W, Dubroka A and Scheuermann R 2008 *Phys. Rev. Lett.* **100** 116601
[4] Lee S L, Kilcoyne S H and Cywinski R (ed.) 1999 *Muon Science*
[5] Blundell S J 2004 *Chem. Rev.* **104** 5717
[6] Patterson B D 1988 *Rev. Mod. Phys.* **60** 69
[7] Meier P F 1982 *Phys. Rev.* A **25** 1287
[8] Anthony J E, Eaton DL and Parkin SR 2002 *Org. Lett.* **4** 15
[9] Anthony J E 2006 *Chem. Rev.* **106** 5028
[10] Pratt F L 2000 *Physica* B **289** 710
[11] Lord J S 2006 *Physica* B **374** 472

[12] Schulz L, Willis M, Nuccio L, Shusharov P, Fratini S, Pratt F L, Gillin W, Kreouzis T, Heeney M, Stingelin N, Stafford C, Beesley D, Bernhard C, Anthony J E, McKenzie I, Lord J S and Drew A J, submitted. arXiv:1006.2651v1

Two dimensional electron gas confined over a spherical surface: Magnetic moment

A Hernando[1], P Crespo[1] and M A García[2]

[1]Instituto de Magnetismo Aplicado, UCM-CSIC-ADIF. Las Rozas. P. O. Box 155, Madrid 28230 and Dpto. Física de Materiales, Universidad Complutense.

[2]Instituto de Cerámica y Vidrio. CSIC c/Kelsen,5 Madrid 28049 .

E-mail: antonio.hernando@adif.es

Abstract. Magnetism of capped nanoparticles , NPs, of non-magnetic substances as Au and ZnO is briefly reviewed. The source of the magnetization is discussed on the light of recent X-ray magnetic circular dichroism experiments. As magnetic dichroism analysis has pointed out impurity atoms bonded to the surface act as donor or acceptor of electrons that occupy the surface states. It is proposed that mesoscopic collective orbital magnetic moments induced at the surface states can account for the experimental magnetism characteristic of these nanoparticles. The total magnetic moment of the surface originated at the unfilled Fermi level can reach values as large as 10^2 or 10^3 Bohr magnetons.

1. Introduction: Magnetic properties of Au and ZnO capped NPs

In this article the characteristics of the magnetism observed in capped Au and ZnO nanoparticles is briefly reviewed. It is now well known from EXAFS analysis [1] that this anomalous magnetism is originated by the modifications of the electronic structure of the surface introduced by the capping molecules. X-ray magnetic circular dichroism, XMCD, studies have pointed out the magnetic moment determined by magnetometry actually lies in the atoms forming the NPs and is not only originated by small amounts of impurities, as the magnitude of the magnetic moment could suggest. The most important characteristics of the observed magnetic behaviour can be summarised as follows: i) in all cases the saturation magnetization that is very low, typically in the range of 10^{-1} or 10^{-2} Bohr magnetons per atom, is reached at moderate applied fields (0.5T) and ii) the magnetization values are weakly dependent of temperature in the range from 4K up to room temperature.

In some cases the magnetization curve presents hysteresis, even though generally with moderate or low coercive field and remanence. Therefore the two main question to solve were: i) the source of the magnetic moment and ii) the origin of the hysteresis, i. e. the origin of possible exchange and anisotropy.

It was found that an alteration of the electronic structure of Au NPs induced by capping with organic molecules also leads to the appearance of room temperature ferromagnetism [1]. Au NPs of 1.6 nm radius were linked to organic molecules via thiol bonds. The NPs assembly exhibited magnetization loop and ferromagnetism like response at room temperature. Even though the magnetic moment per atom is very low, typically in the range between 0.1 to 0.02 Bohr magnetons, and, therefore, is of the order of possible ferromagnetic impurities [2], subsequent X-ray magnetic circular dichroism, appropriately tuned to Au atoms [3,4], as well as Au Mössbauer spectroscopy indicated that the Au atom posses a magnetic moment [4]. Since the first XANES and EXAFS studies it was assumed that the source of the magnetism had to be linked to electrons or holes introduced by the thiol or phosphine ligands between the surface atoms of the NPs and those located at the end of the organic chain. The magnetic moment was thought of as due to uncompensated angular moment or spins of those electrons or holes introduced by the ligands. However, the observed saturation reached at room temperature at relative low fields (1T) and the shape of the magnetization curve represented a great challenge. Even though the existence of a permanent magnetic moment were assumed, the NPs size was small enough to expect a superparamagnetic behaviour unless the magnetic anisotropy were extremely large.

Based on the magnetic behaviour of capped Au NPs we thought that capped semiconductor oxide NPs could exhibit a similar behaviour. Then the magnetic response of capped ZnO NPs was studied. The importance of the subjects is outlined below

Most of technological devices are based on the properties of semiconductors and magnetic materials. Materials able to join in a single one high velocity of processing information, as semiconductors, and information storage, as ferromagnets, should be desirable. Since the proposal by Ohno [5], extensive research work on dilute magnetic semiconductors materials have led to discover traits of ferromagnetism for semiconductors oxide compounds with and without magnetic impurities [6-14]. Despite some initial promising results in Mn:ZnO[10] it is not clear if diluted magnetic semiconductors can show this high temperature ferromagnetism. Hysteresis loops have been also observed in oxide nanoparticles capped with different type of molecules [9,11]. In particular, Kittisltved et al [9] have shown that Mn:ZnO nanoparticles and thin films only exhibit room temperature ferromagnetism when capped with molecules that introduce p-type defects, whereas other capping that introduce n-type defects leads to no room temperature ferromagnetism. On the contrary for Co:ZnO films the n-type defects favour the appearance of ferromagnetism. Several class of defects as O vacancy, cationic vacancy, adatoms or doping have been invoked and speculatively suggested as sources of local magnetic moments, even in the absence of magnetic ions [7,11]. These magnetic moments would be expected to contribute to macroscopic paramagnetism. However, in many cases the samples exhibit low coercivity and low remanence hysteresis loop that seems to be weakly dependent on temperature. Since, the exchange coupling among these local moments has not been satisfactory explained [12-14] the question deals with the possible origin of such traits of ferromagnetic behaviour.

Recently some relevant experiments and calculations were carried out on ZnO, nanoparticles, NPs, [11,15] capped with different molecules. M. A. García et al [11] have shown that 10nm sized ZnO NPs, without doping with magnetic impurities, can exhibit room temperature ferromagnetism when capped with three different organic molecules, amine, thiol and topo. It was also observed that even though no differences in the structure, as determined by X-ray diffraction and transmission electron microscopy, were detected among the three type of nanoparticles, the magnetization correlated with the differences of the electronic configuration of the surface as measured by XANES. This result pointed out that traits of ferromagnetism can be observed in semiconductors NPs and perhaps films and surfaces without introducing magnetic *d* transition atoms as impurities.

Later circular magnetic dichroism experiments were performed by Chaboy et al [15] on thiol capped ZnO NPs. The results clearly indicated that the valence band formed by the Zn-3d and the O-2p overlapping bands is filled and consequently magnetically depolarised. However, the conduction band, mainly formed by empty 4sp Zn surface states, exhibits a clear magnetically polarised signal that enables to conclude that the observed magnetism is originated by 4sp electrons associated with the bonding of the capping molecules to the NP surface atoms. It is interesting to remark that polarization in Au capped NPs was observed in the 5d level but the metallic character of Au and the overlapping between bands does not allow a so clear observation as that obtained in semiconductors. First principle calculations performed by Botello-Méndez et al [16] for passivated nano-ribbons and clusters of ZnO also concluded that bonding to sulphur atoms induces a magnetic moment that could be responsible of the observed ferromagnetic responses.

In order to ascertain the modifications introduced on the surface electronic structure by capping molecules, it is worth to remark that the bonding of the carbon oxide, CO, molecule to the surface zinc ions in ZnO was studied in detail by Anderson and Nichols in 1986. They found that the effect of the complex bonding, that involves the decay to an empty low energy level of the electron that would occupy the anti-bonding state, can be summarised as the donation of one electron from the molecule to the 4sp surface orbitals of Zn [17] that are initially empty. Therefore, the polarisation of the 4sp electrons observed by XMCD should be that of the electrons donated by the capping molecule (or surface defects in general) which are confined in a thin depth at the surface. Hence, the origin of magnetization in capped ZnO NPs must be closely related to those electrons donated by the atoms of the organic molecule that are bound to the Zn atoms of the NP surface. Consequently the relevant issue to address should be the calculation of the magnetic moment of a two dimensional electron gas confined in a spherical surface.

We outline the idea proposed by Hernando and García [18] to explain the origin of the magnetic moment in capped ZnO NPs. First the energy spectrum of the electrons confined on a spherical surface is discussed. It will be shown that for an assembly of NPs, fluctuations of NPs size as well as fluctuations in the coverage factor induce fluctuations in both Fermi level and its occupation number. For partially occupied Fermi level a permanent magnetic moment develops similarly to that induced at incomplete atomic orbitals. However, for the case of NPs the strength of the magnetic moment may be two or three order of magnitude larger than for atoms. After analysing the relative strength of the different energy terms the magnetic moment of the partially filled 4sp band of surface states is estimated [18]. Subsequently, the cause of ferromagnetic traits is discussed by taking into account geometrical deviations from the spherical symmetry. Notice that the results can be extended to any type of NPs with its band of surface states partially occupied by either electrons or holes. Finally the results predict a magnetic moment behaviour that is compared with the more important experimental characteristics of the hysteresis loop observed in capped oxide NPs.

Even though the basic idea of our proposal [18] here outlined was focused to account for the magnetism of capped oxide NPs it might be probably also valid for Au NPs. The clear separation and lack of overlapping of *d* and *sp* bands due to the semiconductor character of ZnO enables the observation of a well defined XMCD polarisation in the sp band[15]. However, XMCD polarisation observed in Au has been restricted to the accessible *d* band [3,4] that overlaps with the *p* one.

2. Energy spectrum of an electron gas confined on spherical surface.

In order to estimate the permanent magnetic moment of the electrons occupying the 4sp surface states we must begin by calculating its energy spectrum and eigenstates. Let us initially consider that the electrons donated by the capping organic molecules are confined by a Dirac δ potential to jump between surface atomic sites thus forming an electron gas depicted by its effective mass. The corresponding eigenstates of a non interacting gas confined in the spherical surface of NPs [19-23] are the spherical harmonics $Y_{l,m}(\theta,\phi)$ and their associated kinetics energy eigenvalues depend on the angular quantum number l as

$$e_l = \frac{l(l+1)\hbar^2}{2mr^2} = \frac{l(l+1)}{2} e_1 \qquad (1)$$

where r is the radius of the NPs and m is the effective mass of the electron, that can be written as $m = \dfrac{\hbar^2}{2a^2 b}$, b being the hopping amplitude proportional to the surface band width and a the average inter-atomic distance at the surface. The experimentally measured width of surface bands varies between 0.5 and 1.0 eV [24]. Notice that e_l can be expressed as $e_l = \dfrac{\hbar^2}{2mr^2} = b\left(\dfrac{a}{r}\right)^2$ The distance, d_l, between adjacent levels with quantum numbers l and $l+1$ is given by

$$d_l = (l+1)e_l = (l+1)b\left(\frac{a}{r}\right)^2 \qquad (2)$$

Consider the surface to be formed by N atoms and that there are N_b ($<N$) electrons, donated by $N_b=\alpha N$ capping molecules (or defects) with $\alpha<1$, where N is of the order of $4\pi\left(\frac{r}{a}\right)^2$. If we proceed as in the case of free electrons by filling successively the levels of increasing energy given by (1) up to reach N_b and after taking into account that the maximum number of electrons that can occupy any level is given by $2(2l+1)$ it can be seen that the Fermi level is implicitly defined through the following relationship

$$N_b = \alpha 4\pi\left(\frac{r}{a}\right)^2 = 2(l_F+1)^2 \tag{3}$$

In general l_F as defined by (3) is not an integer but the following relation can always be written

$$l_F = \sqrt{\frac{N_b}{2}} - 1 = q + x, \tag{4}$$

where q is an integer and $0<x<1$. The Fermi level, is then given by: $l_F = q+1$, and must be only partially filled. According to (3), the number of electrons occupying the first q levels that are completely filled is $2(q+1)^2$. Therefore, the number, n^*, of electrons occupying the Fermi level should be $n^*= N_b-2(q+1)^2$, that according to (4) becomes

$$n^*= 2x(2q+2+x)=2x(2l_F+x) \tag{5}$$

Notice that the band would be filled for $\alpha=2$. As any bonding molecule contributes with an electron to the surface band the maximum possible number of electrons that is reached for a coverage factor $\alpha=1$ would correspond to a half filled band. Consequently, after capping, the surface band is generally populated less than one half of its capacity. If we restrict our analysis for NPs ranging in diameter between 1 to 20 nm, (r/a) roughly varies between 5 and 100 that correspond to N comprised in the interval 300 to 30000, respectively. Therefore, for the maximum capping coverage ($\alpha=1$), l_F would be comprised between 12 and 123.

3. Intra-band exchange interaction and spin-orbit coupling

Besides the kinetic energy the surface electrons are subjected to electron-electron interactions as well as to surface spin-orbit coupling. The relative strength of these Hamiltonian terms would determine, as it happens in atomic physics, the order of perturbation steps to approximate the wave function. In other words, the spontaneous magnetic moment of the N_b electrons of a single NP is expected to be that associated with the n^* electrons occupying its partially filled degenerated Fermi level. However, even if n^* is known the estimation of the moment requires also to know the way in which these individual moments are coupled to create the collective moment of the surface state.

When the kinetic energy is assumed to be the main contributing term to the total Hamiltonian the spherical harmonics form the complete set of eigenfunctions in which the perturbation process should be carried out. However, two limit cases of angular momentum coupling should be considered as is well known from atomic physics. For the case of electron–electron interaction larger than the spin-orbit coupling, the angular moments couple to each other according to the *L-S* coupling. Under opposite conditions the coupling becomes *j-j* type. Independently of the type of coupling the values of the total angular momentum quantum number, J, is the same in both cases even though the corresponding energies can differ in absolute value. As we are here only concerned with the magnetic moment the calculation will be restricted to the case of *L-S* coupling. For the *L-S* coupling the progressive filling of the Fermi level proceeds according to the Hund rules, i. e. maximum S and maximum allowed L. But is important to take into consideration that the estimation of the magnetic moments, obtained from the estimation of the possible J values at the Fermi level, is independent of whether the exchange is larger or smaller than the spin-orbit coupling, provided that the distance between adjacent kinetic energy levels is larger than the spin-orbit coupling.

The importance of the spin-orbit coupling at surfaces (Rashba effect), $\gamma l\sigma$, where γ is the Rashba coefficient σ the Pauli matrices, has been experimentally pointed out by many authors that measured and calculated its splitting [24-30].

4. Calculation of the permanent magnetic moment associated with the Fermi Level

As previously stated we can calculate the possible values of J in the Fermi level, characterised by l_F and occupied by $n^*<2(2l_F+1)$ by using either L-S or j-j coupling. Let us consider L-S or Russel-Saunders coupling. Then according to the Hund rules the electrons occupy successively the Fermi level by reaching the maximum spin and the maximum orbital momentum compatible with the maximum spin condition. The values of L and S are symmetric respect to $n^*=(2l_F+1)$. Therefore, the following relationship holds for L and S, $L(n^*) = L(2(2l_F+1)-n^*)$; n^* can then be restricted for calculation to the condition $0<n^*<(2l_F+1)$.

Then for L-S coupling, if the spectral term with lower energy is that with quantum numbers $L(n^*)$ and $S(n^*)$ the value of $J(n^*)$ corresponding the ground state is either $J= L+S$ or $J= L-S$ for negative or positive sign of spin-orbit coupling, respectively. According to the Hund rules the spectral term obtained by considering n^* electrons in the l_F level is that given by $L=\left(l_F-\frac{n^*-1}{2}\right)n^*$. Therefore, for negative spin-orbit coupling the value of $J(n^*)$ becomes

$$J(l_F, n^*)=\left(l_F+1-\left(\frac{n^*}{2}\right)\right)n^* \quad (6)$$

whereas for positive spin-orbit coupling, $J(l_F, n^*)$ can be written as

$$J(l_F, n^*)=\left(l_F-\left(\frac{n^*}{2}\right)\right)n^* \quad (7)$$

In general the spin-orbit coupling is negative for orbital level occupied below one half of the total occupation and positive for higher occupation numbers.

Figure 1 illustrates the J dependence on n^* for the typical size of NPs reported in the literature and for negative sign of spin-orbit coupling. It can be seen that the values of J (l_F, n^*) for l_F*=8,20,40 and 80* are very large when compared to the usual strength of the atomic angular moments.

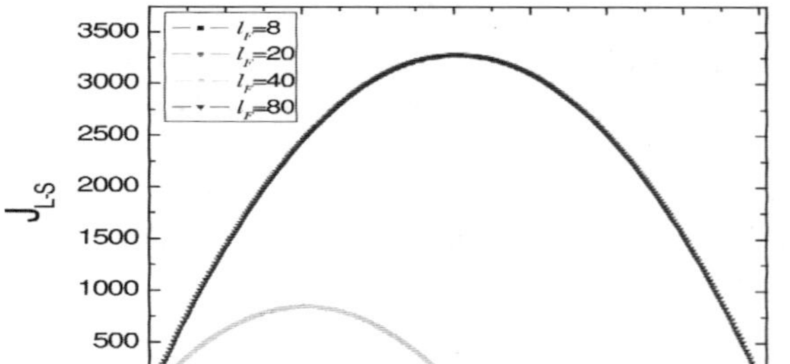

Figure 1. J values as a function of n^* obtained according to (4) for l_F ranking between 8 y 90, that are the expected values for normal NPs sizes as indicated at page 4. The average values $<J>$ obtained for uniform n^* distribution and obtained from (7) are, 21, 133, 533 and 2133 for $l_{F=}$, 8, 20, 40 and 80 respectively.

5. The permanent magnetic moment of an assembly of NPs.

In a NPs assembly both $\frac{r}{a}$ and α shall present fluctuations respect its average values $\left\langle\frac{r}{a}\right\rangle$ and $\langle\alpha\rangle$ that will give rise, according to (3), to l_F and n^* fluctuations. In fact the following approximately relationship is easily derived from (3)

$$\frac{\Delta l_F}{\langle l_f\rangle} \approx \frac{\Delta r}{\langle r\rangle} + \frac{\Delta\alpha}{\langle\alpha\rangle} \tag{8}$$

Notice that, for $\langle l_f\rangle$=100, the relative fluctuation in radius that leads to a change of Fermi level, Δl_F =1, is given by 0.01. Therefore small fluctuations respect to the experimental ones must give rise to a wide spectrum of Fermi levels with occupation, n^*, ranging continuously between 1 and $2(2l_F+1)$. For a constant l_F and a uniform distribution of n^* the average value of the quantum magnetic moment number, also indicated in Figure 1 caption, becomes:

$$<\mu>_{l_F} = \mu_B < g_J J(l_F, n^*) > \tag{9}$$

where g_J is the Landé factor for the particular $2J+1$ subspace corresponding to the quantum number J, i. e. $g_J = 1 + \frac{j(j+1) - L(L+1) + S(S+1)}{2J(J+1)}$.

The order of magnitude of the average $<\mu>_{l_F}$ when n^* varies between 1 and $2(2l_F+1)$ can be estimated from the analytical expressions (6) or (7), after approximating[32] $g_J=1$. For n^* larger than $2(l_F+1)$, negative spin-orbit coupling, (6) yields

$$<\mu>_{l_F} = \frac{1}{3}(l^2{}_F + 2l_F + 1)\,\mu_B \tag{10}$$

whereas, for n^* smaller than $2(l_F+1)$, positive spin-orbit coupling, (7) yields

$$<\mu>_{l_F} = \frac{1}{3}(l^2{}_F - \frac{1}{2}l_F + \frac{1}{2})\ \mu_B \tag{11}$$

$<\mu>_{l_F}$ expressed by (10) and (11) is the average magnetic moment of the NPs assembly for a constant l_F and a uniform n^* distribution and is given by

$$<\mu>_{l_F} = \frac{1}{3}(l^2{}_F + \frac{1}{4}l_F + \frac{1}{4})\ \mu_B \tag{12}$$

From (10) and (11) it becomes evident that for large l_F, i. e, l_F>10 , the order of magnitude of the average magnetic moment at l_F is given by $\mu_B l^2{}_F$.

It is remarkable to observe that the nanometric scale reflected in the ratio $\frac{a}{r}$ governs, i) the importance of the kinetic energy as it enhances d_l in (2) and ii) the value of the average magnetic moment per atom that being originated exclusively at the surface of the NPs becomes macroscopically negligible when the ratio $\frac{a}{r}$ becomes of the order of the sensitivity and precision degree of the magnetometry instrument.

6. Effect of the applied magnetic field

When a magnetic field is applied along the polar or z axis of the NPs the spherical symmetry of the Hamiltonian is partially broken by the field. As thoroughly analysed by Huang and Gumbs [23], in the low field range the spectrum of the kinetic energy only is modified by a perturbation that splits the degenerated ground state in its $2J+1$ components and induces a diamagnetic moment given by

$N_b \dfrac{e^2 B\, r^2}{4m}$ that, even though is much larger that that observed for atoms as a consequence of the mesoscopic scale of the orbits[33], it remains almost negligible for $J \neq 0$, since, for B=1T, it is of the order of $10^{-3}\mu_B N_b$ (low field approximation in refs [22] and [23]). It is to be remarked that enhancement of the orbital moment at plane surfaces, associated with the surface broken symmetry, has been observed and reported by different authors [34,37].

According to our calculations the NPs should exhibit giant paramagnetism; reaching saturation under moderated fields even at room temperature. Notice that with a surface collective magnetic moments of the order of $g\mu_B l_F^2$, (8), an applied magnetic field of 1T would induce a maximum Zeeman splitting of $2\mu_B l_F^2$ that is larger than the thermal energy at room temperature for any l_F>10. It can be concluded that the Langevin or Brillouin curve depicting the paramagnetism of the NPs is close to saturation, i.e. , $\dfrac{\mu_B\left(\frac{1}{3} l_F^{\,2}\right)B}{kT} \geq 1$, for B= 1.5T and T= 300K if l_F is larger than 10, that is a very reasonable value for the size of the NPs experimentally studied, and coverage factor of the order of 0.5. The experimental saturating field is around 1.5 T .

7. Conclusions

In conclusion it has been shown that NPs with radius below 10 nm, even being of diamagnetic substances, can exhibit giant paramagnetism when capped with molecules that act as donors or acceptor of surface electrons. The experimental values are very well reproduced in the framework we have reported here. The magnetic saturation, typically of 0.1 to 0.01 Bohr magneton per NPs atom, is reached under moderated fields at room temperature, in the range between 1 and 2 T, due to the large strength of the collective magnetic moment of the NP, generally larger than 100 Bohr magnetons . The source of the magnetic moments is the two dimensional electron (or hole) gas introduced by impurities, as for instance capping molecules, that occupy previously empty surface states (or introduce holes in a filled band). Orbital and spin angular moments are formed at the generally unfilled Fermi level of the system that is confined on a spherical surface. The unfilled Fermi level of the surface band is characterised by a collective magnetic moment just like an atomic orbital but with much larger quantum number.

The effect of confinement at the NPs surfaces of the electrons actives in the bonds to the capping molecules together with the unfilled character of its Fermi level are found to be the source of the exotic magnetism of these non-magnetic in bulk, NPs.

Acknowledgments

This work has been supported by the Spanish Government through projectsMAT2009-14741-C02-01 and CSD2007-00010

References

[1] Crespo P et al. 2004 *Phys. Rev.Lett.* **93** 8.
[2] Crespo P et al. 2008 *Phys. Rev. Lett.* **97** 177203.
[3] García M A et al. 2009 *J. Nanosci.Nanotecnol.* **9**-6434.
[4] Garitaonanindía J et al. 2008 *Nano Lett.* **8** 661.
[5] Ohno H 1998 *Science* **281** 951; Dietl T et al. 2000 *Science* **287** 1019.
[6] Yoon S D et al. 2006 *J. Phys. Condens. Matter* **18** L355-L361.
[7] Sundaresan A et al. 2006 *Phys. Rev.B* **74** 161306; Venkatesan M, Fitgerald C B and Coey J M D 2004 *Nature* **430** 630; Venkatesan M, Fitzgerald C B, Lunney J G, and Coey J M D,

2004 *Phys. Rev. Lett.* **93** 177206.
[8] Afanasev V V and Stesmans A 2004 *J. Appl. Phys.* **95** 2518.
[9] Kittilstved K R, Norberg M S and Gamelin D R. 2005 *Phys. Rev. Lett.* **94** 147209.
[10] Sharma P et al. 2003 *Nat. Mat* **2** 673; García M A et al. 2005 *Phys. Rev. Lett.* **94** 217206.
[11] García M A et al. 2007 *Nano Lett.* **7** 1489.
[12] Elfimov I S et al. 2002 *Phys. Rev. Lett.* **89** 216403.
[13] Bouzerar G and Ziman T, 2006 *Phys. Rev. Lett.* **96** 207602. Recent calculations investigate further possible exchange mechanism as superexchange in dilute dielectrics as in Kikoin K and Fleurov V 2006 *Phys. Rev. B* **74** 174407.
[14] Osorio-Guillén J et al. 2006 *Phys. Rev. Lett.* **96** 107203. Jun Chen et al. 2006 *Phys. Rev. B* **74** 235207.
[15] Chaboy J et al. 2010 *Phys. Rev. B.* **82** 064411.
[16] Botello-Méndez A R 2008 et al. *Nano Res* **1** 420.
[17] Anderson A B and Nichols J A 1986 *J. Am. Chem. Soc.* **108** 1385.
[18] Hernando A and García M A 2010 55th MMM Atlanta and NANO 2010 (Rome).
[19] Chakraborty T and Pietiläinen P 1994 *Phys. Rev. B.* **50** 8460.
[20] Fuhrer A et al. 2001 *Nature* **413** 822.
[21] Rumpf K, Granitzer P and Krenn H, 2008 *J. Phys. Condens. Matter.* **20** 454221.
[22] Tempere J, Silvera I F and Devreese J T 2002 *Phys. Rev. B* **65** 195418.
[23] Huang D and Gumbs G 1992 *Phys. Lett. A* **167** 396.
[24] Lashell S et al. 1996 *Phys. Rev. Lett.* **77** 3419.
[25] Haverkort M H et al. 2008 *Phys. Rev. Lett.* **101** 026406.
[26] Hoesch M et al, 2004 *Phys. Rev. B* **69** 241401.
[27] Hochstrasser M et al. 2002 *Phys. Rev. Lett.* **89** 216802.
[28] Rotenberg E, Chung J W and Kevan S D 1999 *Phys. Rev. Lett.* **82** 4066.
[29] Sih V et al. 2005 *Nature Phys.* **1** 31.
[30] Meier L et al. 2007 *Nature Phys.* **3** 650.
[31] Hernando A, Crespo P and García M A, 2006 *Phys. Rev. Lett.* **96** 057206; Hernando A et al. 1997 *Phys. Rev. B* **56** 7800.
[32] The value of g_J varies between 1 for pure orbital moment and 2 for pure spin moment. Therefore, it has to be averaged in the Fermi level since different L and S corresponds to each n^*. For the sake of simplicity we have considered it equal to one.
[33] Hernando A et al. 1997 *Phys. Rev. B* **56** 7800.
[34] Liu C, Matsuda I, Hobara R and Hasegawa S 2006 *Phys. Rev. Lett.* **96** 036803.
[35] Meier L et al 2007 Sept. *Nature Phys.* **3** 650.
[36] Van der Laan G et al. 1992 *Phys. Rev. Lett.* **69** 3827.
[37] Weller D et al. 1995 *Phys. Rev. Lett.* **75** 3572.

Journal of Physics: Conference Series **292** (2011) 012006 doi:10.1088/1742-6596/292/1/012006

Spin torque and critical currents for magnetic vortex nano-oscillator in nanopillars

K Y Guslienko[1,2], G R Aranda[3] and J Gonzalez[1]

[1] Dpto. Fisica de Materiales, Universidad del Pais Vasco, 20018 Donostia-San Sebastian, Spain
[2] IKERBASQUE, the Basque Foundation for Science, 48011 Bilbao, Spain
[3] Centro de Física de Materiales UPV/EHU-CSIC, 20018 San Sebastián, Spain

E-mail: sckguslk@ehu.es

Abstract. We calculated the main dynamic parameters of the spin polarized current induced magnetic vortex oscillations in nanopillars, such as the range of current density, where vortex steady oscillations exist, the oscillation frequency and orbit radius. We accounted for both the non-linear vortex frequency and non-linear vortex damping. To describe the vortex excitations by the spin polarized current we used a generalized Thiele approach to motion of the vortex core as a collective coordinate. All the calculation results are represented via the free layer sizes, saturation magnetization, and the Gilbert damping. Predictions of the developed model can be checked experimentally.

1. Introduction

Now excitations of the microwave (mw) oscillations in magnetic nanopillars, nanocontacts and tunnel junctions by spin polarized current as well as the current induced domain wall motions in nanowires are one of perspective applications of spintronics [1]. A general theoretical approach to mw generation in nanopillars/nanocontacts driven by spin-polarized current based on the universal model of a non-linear auto-oscillator was developed recently [2]. The model was applied to the case of a spin-torque oscillator (STO) excited in a uniformly magnetized free layer of nanopillar, and explains the main experimentally observed effects such as the power and frequency of the generated mw signal. However, the low generated power ~ 1 nW of such STO prevents their practical applications. Recently extremely narrow linewidth of 0.3 MHz and a relatively high generated power was detected for the magnetic vortex (strongly non-uniform state) nano-oscillators in nanopillars [3]. The considerable mw power emission from a vortex STO in the MgO-based magnetic tunnel junctions was observed [4]. It was established that the permanent perpendicular to the plane (CPP) spin polarized current I can excite magnetic vortex motion in free layer of the nanopillar if the current intensity exceeds some critical value, I_{c1} [5]. Then, in the interval $I_{c1} < I < I_{c2}$ (I_{c2} is a second critical current) there is mw generation at a frequency which smoothly increases with the current increasing [6]. This frequency corresponds to the vortex oscillations with a stationary orbit determined by the current value, I. If the current exceeds I_{c2} [7,8], the vortex steady motion state is not stable anymore, presumably because the vortex reaches a critical velocity [9] and reverses its core. The vortex with opposite core polarization cannot be excited for the given current sign I and the oscillations stop. Such excitation scheme is different from the current-in-plane (CIP) case, where one needs to apply a.c. CIP of about the resonance frequency to excite the vortex motion [10]. Low value of I_{c1} and high value of I_{c2} make the vortex CPP nano-oscillators attractive for applications as microwave devices. However, the calculations of the spin torque (ST force) gave contradictory results for the ST force and I_{c1}. The standard STO approach [2] is not applicable to the vortex dynamics due to a specific vortex damping term. The problem was reduced to the problem of vortex core reversal in the perpendicular to the layer plane magnetic field [7] but the vortex steady state dynamics was not accounted. Using the Thiele approach the ST force calculated in Refs. 5, 8 is in 2 times bigger than one calculated from the energy dissipation balance [6]. Non-linearity of the main governing parameters (vortex frequency and damping) and the Oersted field of current were not accounted or accounted incorrectly. The physical mechanism responsible for the critical current I_{c2} has not yet been clarified and the value of I_{c2} has not been calculated.

2. Calculation results and discussion

We present below a simple approach to calculate the ST force and the critical currents of the vortex STO in nanopillar. The approach is based on the Thiele's formulation of the non-uniform magnetization dynamics [11] and conception of the linear spin excitations [12]. The nanopillar device consists of two ferromagnetic layers, typically FeNi or Co and a nonmagnetic metallic spacer, typically Cu, arranged in a vertical stack (figure 1). Magnetization of one layer is fixed (this layer is so called polarizer), whereas the magnetization $\mathbf{M}(\mathbf{r},t)$ of the second layer of the nanopillar is free to rotate. The current of spin polarized electrons transfers some torque $\boldsymbol{\tau}_s$ from the polarizer, which excites magnetization dynamics of the free layer. We start from the Landau-Lifshitz equation of motion $\dot{\mathbf{m}} = -\gamma \mathbf{m} \times \mathbf{H}_{eff} + \alpha_{LLG} \mathbf{m} \times \dot{\mathbf{m}} + \gamma \boldsymbol{\tau}_s$, where $\mathbf{m} = \mathbf{M}/M_s$, M_s is the saturation magnetization, $\gamma>0$ is the gyromagnetic ratio, $\mathbf{H}_{\rm eff}$ is the effective field, and α_{LLG} is the Gilbert damping. We use the ST term in the form suggested by Slonczewski [13], $\boldsymbol{\tau}_s = \sigma J \mathbf{m} \times (\mathbf{m} \times \mathbf{P})$, where $\sigma = \hbar\eta/(2|e|LM_s)$, η is the current spin polarization (η=0.2 for FeNi), e is the electron charge, L is the free layer (dot) thickness, J is the current density, and $\mathbf{P} = P\mathbf{z}$ is the unit vector of the polarizer magnetization (P=+1/-1). We assume the positive vortex core polarization (direction of perpendicular to the dot plane magnetization) p=+1, P=+1 and define the current (flow of the positive charges) as positive I>0 when it flows from the polarizer to free layer. The spin polarized current can excite a vortex motion in the free layer if only $IpP > 0$ (only the electrons bringing a magnetic moment from the polarizer to free layer opposite to the core polarization can excite a vortex motion). Except p, the magnetic vortex is described by its core position, $\mathbf{X}$= (X,Y), and chirality $C=\pm 1$ [14]. Let us denote the Slonczewski´s energy density of the spin polarized current as w_s. Then, using the Thiele approach and the ST effective field $\partial w_s / \partial \mathbf{m} = a\mathbf{m} \times \mathbf{P}$, the ST force acting on the vortex in the free layer can be written as

$$F_{ST}^{\alpha} = -\frac{\partial}{\partial X_\alpha} \int dV w_s = aL\mathbf{P} \cdot \int d^2\boldsymbol{\rho} \left(\mathbf{m} \times \frac{\partial \mathbf{m}}{\partial X_\alpha} \right), \tag{1}$$

where $a = M_s \sigma J$, $\alpha = x, y$, $\boldsymbol{\rho} = (\rho, \varphi)$ is the in-plane radius vector, the derivative is taken with respect to the vortex core position $\mathbf{X}$ assuming an ansatz $\mathbf{m}(\boldsymbol{\rho},t) = \mathbf{m}[\boldsymbol{\rho}, \mathbf{X}(t)]$ ($\mathbf{m}$ dependence on the thickness coordinate z is neglected). $|\mathbf{X}|$ has sense of the amplitude of the vortex gyrotropic motion.

We use representation of **m**-components by the spherical angles Θ, Φ (figure 1) as $\mathbf{m} = (\sin\Theta\cos\Phi, \sin\Theta\sin\Phi, \cos\Theta)$ and find the expression for the ST force

$$\mathbf{F}_{ST} = aL \int d^2\boldsymbol{\rho} \sin^2\Theta \frac{\partial \Phi}{\partial \mathbf{X}}. \tag{2}$$

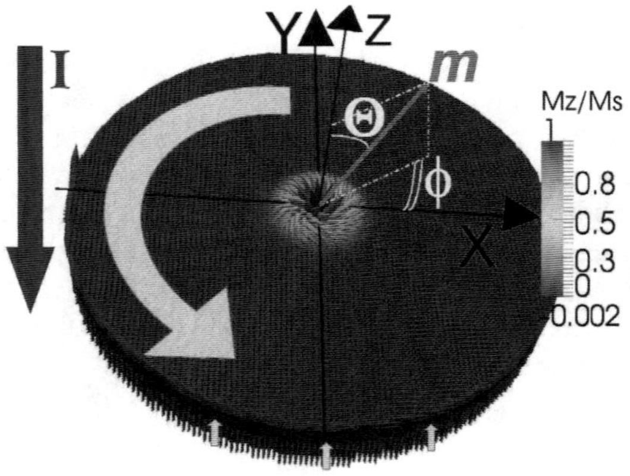

Figure 1. Magnetic nanopillar with the coordinate system used. The upper (free) layer is in the vortex state with the chirality C=+1. The polarizer (red colour) is magnetized uniformly along *Oz* axis. The negative current *I* (vertical arrow) flows from the free layer to the polarizer.

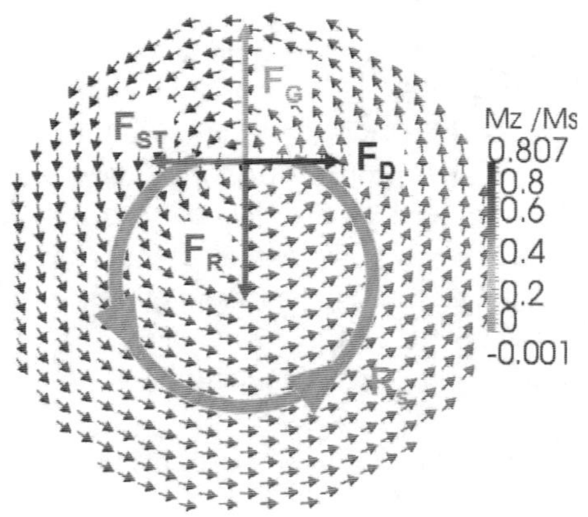

Figure 2. Top view of the free layer with a moving magnetic vortex. The arrows denote a balance for the vortex core of the spin torque ($\mathbf{F}_{ST}$), damping ($\mathbf{F}_D$), restoring ($\mathbf{F}_R$) and gyro- ($\mathbf{F}_G$) forces. The vortex core steady motion radius R_s is marked by orange colour.

In the main approximation we use the decompositions $m_z(\boldsymbol{\rho},\mathbf{X}) = \cos\Theta = m_z^0(\rho) + g(\rho)(\mathbf{X}\cdot\hat{\boldsymbol{\rho}})$, $\Phi(\boldsymbol{\rho},\mathbf{X}) = \Phi_0 + m_0(\rho)[X\sin\varphi - Y\cos\varphi]$, where $m_z^0(\rho) = p(c^2-\rho^2)/(c^2+\rho^2)$, $\Phi_0 = \varphi + C\pi/2$ are the static vortex core profile and phase [14], $g(\rho) = 4pc^2\rho(1+\rho^2)/(c^2+\rho^2)^2$ is the excitation amplitude of the *z*-component of the vortex magnetization ($c = R_c/R$, R_c is the vortex core radius, *ρ*, *X*, *Y* are in units of *R*) and $m_0(\rho) = (1-\rho^2)/\rho$ is the gyrotropic mode profile [12]. One can conclude from Eq. (2), by separating the core and outside-core contributions, that only the moving vortex core contributes to the ST force because the contribution of the main dot area where $\Theta = \pi/2$ is equal to zero due to vanishing integrals of the derivative $\partial_{\mathbf{X}}\Phi$ over the azimuthal angle φ. This is a reason why the ST contribution is relatively small being comparable with the damping contribution.

The integration in Eq. (2) yields the ST force $\mathbf{F}_{ST} = \pi a L(\hat{\mathbf{z}}\times\mathbf{X})$, which is in 2 times smaller than in Refs. 5,8 due to incorrect using of the Thiele's approach there. This force contributes to the Thiele's equation of motion $\mathbf{G}\times\dot{\mathbf{X}} = -\partial_{\mathbf{X}}W + \hat{D}\dot{\mathbf{X}} + \mathbf{F}_{ST}$, where $\mathbf{G} = \hat{\mathbf{z}}2\pi pLM_s/\gamma$ is the gyrovector, $\hat{D}$ is the damping tensor. The vortex energy $W(\mathbf{X})$ and restoring force $\mathbf{F}_R = -\partial_{\mathbf{X}}W$ can be calculated from an appropriate model of the shifted vortex [14] (the force balance is shown in figure 2). For a circular steady state vortex core motion the $\dot{\mathbf{X}} = \boldsymbol{\omega}\times\mathbf{X}$ relation holds, which allows calculating J_{c1}. To calculate the vortex steady orbit radius $R_s = |\mathbf{X}|$ we need to account non-linear on X_α terms in the vortex damping and frequency (the account only non-linear frequency as in Ref. 5 is not sufficient). The gyrovector also depends on **X**, but this dependence is essential only for the vortex core *p* reversal, when **G** changes its sign. The most important non-linearity comes from the damping tensor defined as

$$D_{\alpha\beta}(\mathbf{X}) = -\alpha_{LLG}\frac{M_s}{\gamma}\int dV\,\frac{\partial\mathbf{m}}{\partial X_\alpha}\cdot\frac{\partial\mathbf{m}}{\partial X_\beta}, \tag{3}$$

or $D_{\alpha\beta} = -\alpha_{LLG}(M_s L/\gamma)\int d^2\boldsymbol{\rho}\left[\partial_\alpha\Theta\partial_\beta\Theta + \sin^2\Theta\,\partial_\alpha\Phi\partial_\beta\Phi\right]$ in Θ,Φ-representation. Accounting $D_{\alpha\beta} = D\delta_{\alpha\beta}$ and introducing dimensionless damping parameter $d = -D/|G| > 0$ [15] we can write the equation for a steady state vortex motion with the orbit radius $R_s = |\mathbf{X}|$: $d(s)|G|\omega_G(s)sR = F_{ST}^{\varphi}$ from which $s = R_s / R$ and the critical currents J_{c1}, J_{c2} can be found. In the second order non-linear approximation $d(s) = d_0 + d_1 s^2$, $\omega_G(s) = \omega_0 + \omega_1 s^2$ and $F_{ST}^{\varphi} = \pi a L R s$, where ω_G is the vortex precession frequency, $\omega_1 > 0$ is a function of the dot aspect ratio $\beta = L/R$ calculated from the vortex energy decomposition $W(s)$ in series of $s = |\mathbf{X}|/R$ within the pole free model of the shifted vortex [13]. It can be shown that $\omega_0(\beta) = (20/9)\gamma M_s \beta[1 - 4\beta/3]$ and $\omega_1(\beta) \approx 4\omega_0(\beta)$ for quite wide range of β= 0.01-0.2 of practical interest, whereas considerably larger non-linearity $\omega_1(\beta)/\omega_0(\beta) = 42.8$ was calculated in Ref. 5 due to incorrect account of the magnetostatic energy. We use the pole free model of the shifted vortex $\mathbf{m}[\boldsymbol{\rho}, \mathbf{X}(t)]$, where the dynamic magnetization satisfies the strong pinning boundary condition at the dot circumference [16] $\rho = R$. The damping parameters are $d_0 = \alpha_{LLG}(5/8 + \ln(R/R_c)/2)$, $d_1 = \alpha_{LLG}(R^2/R_c^2 - 8/3)/4$. We need also to account for the Oersted field of the current, which leads to contribution to the vortex frequency proportional to the current density $\omega_0 = \omega_0(\beta, J) = \omega_0(\beta) + \omega_e J$, where $\omega_e = (8\pi/15)(\gamma R/c)\xi C$, $\xi = 1 + 15(\ln 2 - 1/2)R_c/8R$ is the correction to ω_e of Ref. 6 (typically ~10%) for the finite core radius $R_c \ll R$. In the linear approximation we get the equality of the damping force and the ST force (negative damping) as the condition to find the value of J_{c1}, with the contribution of the Oersted field of the current accounted. The values of F_{ST}, J_{c1} coincide with calculation of Ref. 6 conducted by the method of the damping energy balance and differ by the multiplayer 2 from Refs. 5, 8. The first critical current is $J_{c1} = d_0\omega_0 / (\gamma\sigma/2 - d_0\omega_e)$ and the steady state vortex orbit radius is

$$s(J) = \lambda\sqrt{J/J_{c1} - 1}, \quad J > J_{c1}, \quad \lambda^2 = \frac{1}{2}\frac{\gamma\sigma J_{c1}}{[d_1\omega_0(J_{c1}) + d_0\omega_1]}. \qquad (4)$$

In this approximation the vortex orbit radius $s(J)$ increases as square root of the current overcriticality $(J - J_{c1})/J_{c1}$ (for the typical parameters and R=80-120 nm we get λ=0.25-0.30) and the vortex gyrotropic frequency $\omega_G = \omega_0(\beta) + \omega_e J + \lambda^2(J/J_{c1} - 1)\omega_1$ increases linearly with J increasing. The vortex steady orbit can exist until the moving vortex core crosses the dot border s=1 or its velocity $|\dot{\mathbf{X}}|$ reaches the critical velocity υ_c of the core polarization reversal defined in Ref. 9. The vortex with opposite core polarization cannot be excited for the given current sign I and the oscillations stop. The later condition allows to write equation for the second critical current J_{c2} as $\omega_G(J)s(J)R = \upsilon_c$. Substituting to this expression the equations for $\omega_G(J)$ and $s(J)$ derived above we get a cubic equation for J_{c2} in the form $[\gamma\sigma J_{c1}/2d_0 + (\omega_e J_{c1} + \omega_1\lambda^2)x]x^{1/2} = \upsilon_c/\lambda R$, $x = (J_{c2}/J_{c1} - 1)$. This equation has one positive root x_c and the value of J_{c2} can be easily calculated (see figure 3). The former condition (s=1) yields the second critical current $J'_{c2} = (1 + 1/\lambda^2)J_{c1}$. More detailed numerical analysis shows that both the mechanisms of the high current instability of the vortex motion are possible depending on the dot sizes L, R, and the second critical current is the lower value of the calculated currents J_{c2}, J'_{c2}. The moving vortex core reversal inside the dot occurs for large enough R (> 100 nm) and L, whereas the vortex core is expelled from the dot at $J=J'_{c2}$ for small R ($\leq$100 nm) of practical interest for the vortex STO applications in nanopillars. For the typical free layer sizes L=10 nm, R=120 nm and C=1, the critical currents calculated by the above listed equations are J_{c1} =6.3×10^6 A/cm^2 (I_{c1} =2.9 mA), J_{c2} =1.13×10^8 A/cm^2 (I_{c2} =51 mA), and for L= 5 nm, R=100 nm we get J_{c1} =1.8×10^6 A/cm^2 (I_{c1} =0.56 mA), J'_{c2}=2.7×10^7 A/cm^2 (I'_{c2}=8.4 mA).

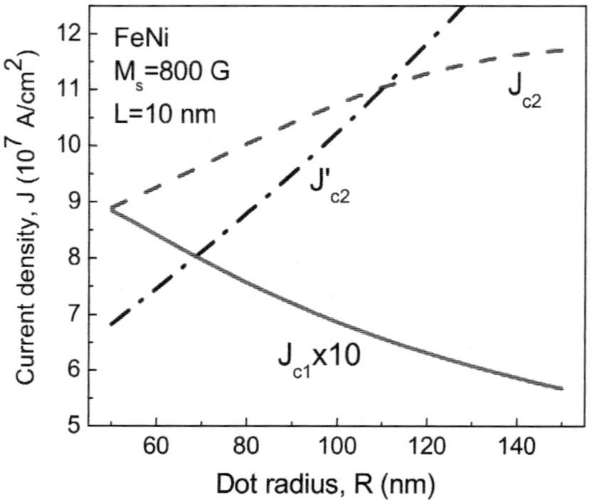

Figure 3. Dependence of the critical currents J_{c1} (solid red line), J_{c2} (dashed green line) and J'_{c2} on the free layer radius R. L= 10 nm, M_s =800 G, η =0.2, $\alpha_{LLG} = 0.01$, $\gamma/2\pi$ = 2.95 MHz/Oe, R_c= 12 nm. The vortex STO motion is stable at $J_{c1} < J < \min(J_{c2}, J'_{c2})$.

3. Conclusions

We calculated the main physical parameters of the spin polarized CPP current induced magnetic vortex oscillations in nanopillars, such as the critical currents J_{c1}, J_{c2}, the vortex steady state oscillations frequency and orbit radius. All the results are represented via the free layer sizes (L, R), saturation magnetization, Gilbert damping and the degree of spin polarization of the current. These parameters can be obtained from independent experiments. The generalized Thiele approach to the dynamics of non-uniform magnetization is applicable to the problem of the vortex STO excitations by the CPP spin polarized current. The spin transfer torque force is related to the vortex core only.

Acknowledgments

The authors thank J. Grollier and A. Khvalkovskiy for fruitful discussions. K.G. and G.A. acknowledge support by IKERBASQUE (the Basque Science Foundation) and by the Program JAE-doc of the CSIC (Spain), respectively. The work was partially supported by the SAIOTEK grant S-PC09UN03. Computation resources and technical support were supplied by SGIKER, CFM and DIPC.

References

[1] Tatara G, Kohno H and Shibata J 2008 *Phys. Rep.* **468** 213
[2] Slavin A, Tiberkevich V 2009 *IEEE Trans. Magn.* **45** 1875
[3] Pribiag VS, Krivorotov IN, Fuchs GD *et al.* 2007 *Nature Phys.* **3** 498
[4] Dussaux A *et al.* 2010 *Nature Commun.* **1** 8
[5] Ivanov BA, Zaspel CE 2007 *Phys. Rev. Lett.* **99** 247208
[6] Khvalkovskiy AV, Grollier J, Dussaux A, Zvezdin AK, Cros V 2009 *Phys. Rev.* B **80** 140401
[7] Caputo JG, Gaididei Y, Mertens FG, and Sheka DD 2007 *Phys. Rev. Lett.* **98** 056604
[8] LiuY, He H and Zhang Z 2007 *Appl. Phys. Lett.* **91** 242501
[9] Guslienko KY, Lee KS, and Kim SK 2008 *Phys. Rev. Lett.* **100** 027203
[10] Kasai S, Nakatani Y, Kobayashi K, Kohno H and Ono T 2006 *Phys. Rev. Lett.* **97** 107204
[11] Thiele AA *Phys. Rev. Lett.* 1973 **30** 230
[12] Guslienko KY, Slavin AN, Tiberkevich V and Kim SK 2008 *Phys. Rev. Lett.* **101** 247203
[13] Slonczewski J 1996 *J. Magn. Magn. Mat.* **159** L1; 2002 *J. Magn. Magn. Mat.* **247** 324
[14] Guslienko KY 2008 *J. Nanosci. Nanotechn.* **8** 2745
[15] Guslienko KY 2006 *Appl. Phys. Lett.* **89** 022510
[16] Guslienko KY, Ivanov BA, Novosad V, Shima H, Otani Y, and Fukamichi K. *2002 J. Appl. Phys.* **91** 8037; Novosad V *et al.* 2005 *Phys. Rev.* B **72** 024455

Critical current density of domain wall oscillation due to spin-transfer torque

Tomohiro Taniguchi[1,2†] **and Hiroshi Imamura**[1*]

[1] Nanosystem Research Institute, National Institute of Advanced Industrial Science and Technology, 1-1-1 Umezono, Tsukuba, Ibaraki 305-8568, Japan
[2] Institute of Applied Physics, University of Tsukuba, 1-1-1 Tennou-dai, Tsukuba, Ibaraki 305-8573, Japan

E-mail: †tomohiro-taniguchi@aist.go.jp, *h-imamura@aist.go.jp

Abstract. The domain wall oscillation due to spin-transfer torque was studied by numerically solving the Landau-Lifshitz-Gilbert (LLG) equation. For a domain wall whose rotation angle $\theta_{\max}$ is less than 180°, we found the existence of the critical current density above which the magnetization dynamics are induced. We studied the dependence of the critical current density on the rotation angle $\theta_{\max}$ and found that the critical current density is proportional to $180^\circ - \theta_{\max}$.

The domain wall oscillation due to spin-transfer torque has drawn great interest because of its potential application to spin-electronics devices such as spin-transfer torque driven nano-oscillator [1, 2]. Franchin *et al.* studied the dependence of the oscillation frequency ν of the 180° domain wall on the electric current density j_e by solving the Landau-Lifshitz-Gilbert (LLG) equation both numerically and analytically, and found that $\nu \propto j_e$ and $\nu \propto j_e^2$ for the small and large current regions, respectively [1]. They also found the static compression of the domain wall where the spatial variation of the magnetization $\mathbf{M}$, $|\partial \mathbf{M}/\partial x|$, is constant in time.

To manipulate the spin-transfer torque driven nano-oscillator, the domain oscillation should be converted to the oscillation of the electric resistance. It is well kwon that the magnetoresistance effect due to a domain wall is proportional to $1/d^2$, where $d \propto 1/|\partial \mathbf{M}/\partial x|$ is the thickness of the wall [3, 4]. Since the compression of the wall is constant in time for 180° wall, as shown by Franchin *et al.*, the oscillation of the 180° wall cannot be detected electrically.

The magnetoresistance effect due to the domain wall in the current-confined-path (CCP) current-perpendicular-to-plane (CPP) spin valve has been studied extensively [5–10]. In CCP-CPP spin valve, the rotation angle of the domain wall is easily controlled by applying the magnetic field [11]. Recently, by solving the LLG equation numerically, Matsushita *et al.* studied the oscillation of the thickness of the domain wall confined in the CCP-CPP spin valve when the rotation angle of the domain wall is less than 180° [2]. Their results open the possibility of the realization of the spin-transfer torque driven nano-oscillator.

In this paper, we study the domain wall oscillation due to spin-transfer torque where the rotation angle of the wall $\theta_{\max}$ is less than 180°. By numerically solving the LLG equation, we found the existence of the critical current density above which the magnetization dynamics in the wall are induced. We study the dependence of the critical current density on the angle $\theta_{\max}$, and found that the critical current density is proportional to $180^\circ - \theta_{\max}$. The dependence of

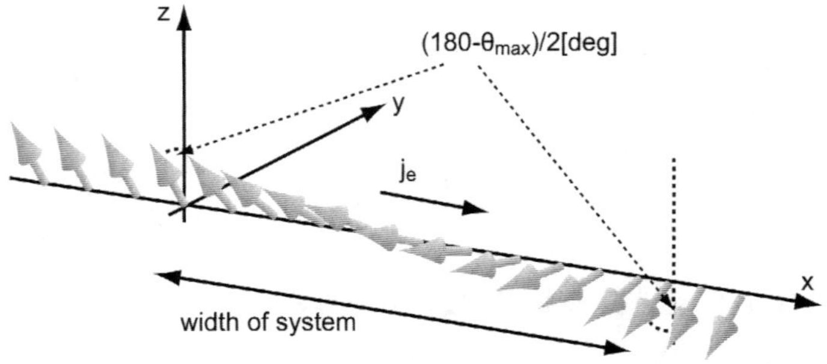

Figure 1. The schematic view of the 1-dimensional domain wall in which the rotation angle of the wall is $\theta_{\max} \leq 180°$. The electric current j_e is applied along to the x-axis.

the critical current density on the Gilbert damping constant and the width of the system are also studied.

We consider the spin-transfer torque driven magnetization dynamics in the one-dimensional domain wall where the rotation angle of the wall is given by $\theta_{\max}$, as shown in Fig. 1. The domain wall lies over $0 \leq x \leq L$, where $L = Na$ is the width of the system, and $N + 1$ and a are the number of the magnetizations in the wall and the lattice constant, respectively. The magnetizations in the wall are coupled via the exchange interaction, and without the electric current, the direction of the k-th magnetization at $x = ka$ ($k = 0 \sim N$) in equilibrium is given by $\mathbf{m}_k = (0, -\sin\theta_k^{(0)}, \cos\theta_k^{(0)})$, where $\theta_k^{(0)} = (180° - \theta_{\max})/2 + (k\theta_{\max}/N)$. It should be noted that the directions of the magnetizations at the boundaries $x = 0, L$, $\mathbf{m}_0$ and $\mathbf{m}_N$, are fixed because there are two magnetic domains in the region $x < 0$ and $x > L$.

By applying the electric current to the domain wall, the spin-transfer from the conduction electrons to the local magnetizations exerts spin-tranfer torque on the magnetizations, which induces the magnetization dynamics in the wall. We assume that the dynamics of the magnetizations in the wall is described by the LLG equation with the spin-transfer torque [12, 13] given by

$$\frac{\partial \mathbf{m}_k}{\partial t} = -\gamma \mathbf{m}_k \times \mathbf{H}_k + b_J \mathbf{m}_k \times \left(\frac{\partial \mathbf{m}_k}{\partial x} \times \mathbf{m}_k \right) - c_J \mathbf{m}_k \times \frac{\partial \mathbf{m}_k}{\partial x} + \alpha \mathbf{m}_k \times \frac{\partial \mathbf{m}_k}{\partial t} , \tag{1}$$

where γ and α are the gyromagnetic ratio and the Gilbert damping constant of the ferromagnetic material, respectively. $\mathbf{H}_k = [2A/(\mu_0 M)]\partial^2 \mathbf{m}_k/\partial x^2$ is the exchange field, where A, μ_0 and M are the exchange coupling constant, the magnetic permeability in vacuum, and the magnetization of the material, respectively. We neglect the demagnetization field whose effect is less important in thinner walls [1]. The strength of the adiabatic and the non-adiabatic spin-transfer torque, b_J and c_J, are given by $b_J = c_J/\xi = \mu_B P j_e/[eM(1 + \xi^2)]$, respectively, where μ_B is the Bohr magneton and P is the spin polarization of the electric current density j_e, respectively. $\xi = 2/(\omega_J \tau_{\rm sf})$ is the ratio of the angular frequency of the spins of the conduction electrons around the exchange field ω_J and the spin-flip scattering probability $\tau_{\rm sf}$ [13].

We solve the LLG equation (1) of the magnetizations in the wall $\mathbf{m}_k$ ($k = 1 \sim N - 1$) numerically. The values of the parameters we used are $A = 1.3 \times 10^{-11}$ J/m, $M = 0.8 \times 10^6$ A/m, $\gamma = 2.21765 \times 10^5$ m/As, $a = 3.0$ Å, $P = 1$, and $\xi = 0.01$, respectively [1].

Figure 2 (a) shows an example of the time evolution of the averaged magnetization $\langle \mathbf{m} \rangle = \sum_{k=1}^{N-1} \mathbf{m}_k/(N-1)$ where $\theta_{\max} = 178°$, $\alpha = 0.02$ and $L = 30.0$ nm, respectively. The value of the electric current density is $j_e = 11.5 \times 10^{10}$ A/m^2. As shown in Fig. 2 (a), the steady precession of the x and y-components of the magnetizations is induced where the precession frequency is 11.9 GHz. The magnetizations in the wall are compressed along the x-direction, as

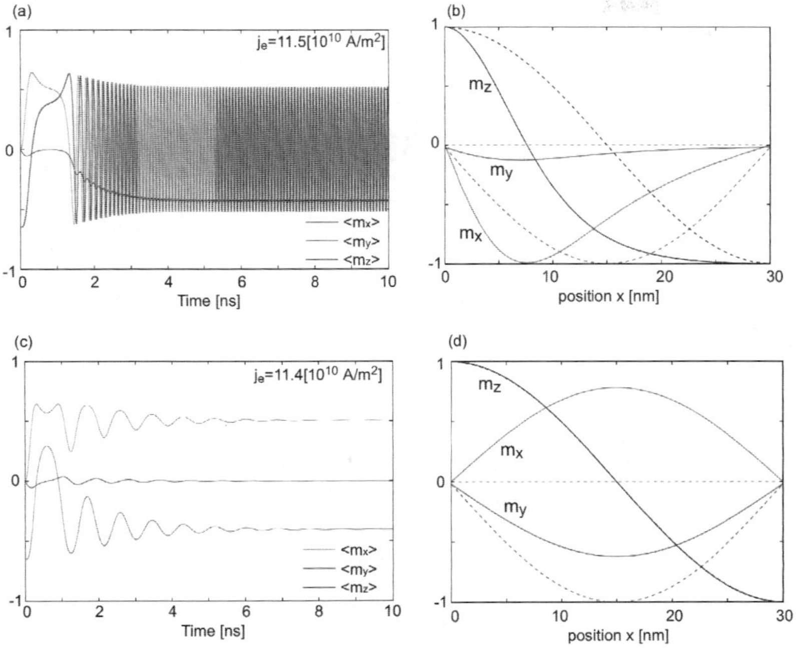

Figure 2. (a) The time evolution of the domain wall dynamics above the critical current density. (b) The magnetization alignment at $t = 0$ (dotted line) and $t = 10$ ns (solid line) above the critical current density. (c) The time evolution of the domain wall dynamics below the critical current density. (d) The magnetization alignment at $t = 0$ (dotted line) and $t = 10$ ns (solid line) below the critical current density.

shown in Fig. 2 (b). These dynamics are similar to those found in Ref. [1]. On the other hand, by decreasing the electric current density, the magnetization dynamics relax and stop in a few nano-second, as shown in Fig. 2 (c), where $j_e = 11.4 \times 10^{10}$ A/m^2. In this case, although the magnetization in the wall first rotate around the z-axis, as shown in Fig. 2 (d), the wall is not compressed. From Figs. 2, it was found that the electric current density should be larger than a certain value (critical current density) to induce the steady precession of the magnetizations. It should be noted that the critical current density of the 180° wall is zero [1].

The physics behind the above results are as follows. Before applying the electric current, the magnetizations inside the wall vary uniformly, i.e., the relative angle between the neighboring magnetizations is constant, $\theta_{\max}/N$, and the exchange energy $\propto -\sum_{i=0}^{N-1} \mathbf{m}_i \cdot \mathbf{m}_{i+1}$ is minimized. By applying the electric current, due to the spin-transfer torque, the x and y-components of the magnetizations try to rotate around the z-axis, as shown in Figs. 2 (d). Then, for $\theta_{\max} < 180°$, the relative angles between the magnetizations at the boundaries ($x = 0, L$) and that inside the wall ($0 < x < L$) increase compared to the equilibrium alignment, which increases the exchange energy of the system. Thus, to induce the magnetization dynamics, the electric current density should be larger than the critical value which is determined by the increase of the exchange energy. This is the origin of the critical current density. On the other hand, when $\theta_{\max} = 180°$, the relative angle, and therefore the exchange energy in the wall is constant during the rotation of the magnetization around the z-axis, and thus, the critical current density is zero.

We studied the dependence of the critical current density on the rotation angle $\theta_{\max}$, the Gilbert damping constant α, and the width of the system L numerically, as shown in Figs. 3 (a), (b), and (c), respectively. From Fig. 3 (a), where $\alpha = 0.02$ and $L = 30$ nm, it is clear that

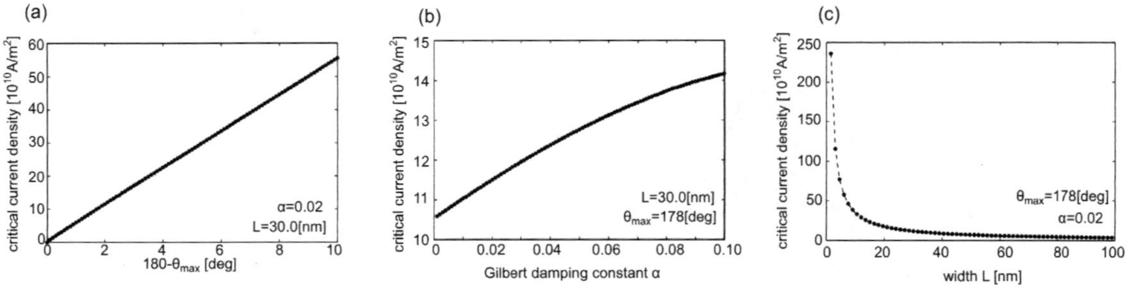

Figure 3. The dependence of the critical current density on (a) the deviation of the rotation angle $\theta_{\max}$ from 180°, (b) the Gilbert damping constant α, and (c) the width of the system L.

the critical current density is proportional to $180° - \theta_{\max}$, and is zero for $\theta_{\max} = 180°$. The value of the current density for $\theta_{\max} = 170°$ is more than 50×10^{10} A/m^2, which is comparable to the current density available in experiments. From Fig. 3 (b), where $\theta_{\max} = 178°$ and $L = 30$ nm, we find that the critical current density is approximately proportional to the Gilbert damping constant α, and remains finite in the limit of $\alpha \to 0$. As shown in Fig. 3 (c), where $\theta_{\max} = 178°$ and $\alpha = 0.02$, the critical current density increases rapidly with decreasing the width of the system L, and reach more than 10^{12} A/m^2 when L is on the order of a few nano-meter. Since this value is much higher than the electric current density available in experiments, we cannot expect to observe the domain wall oscillation for a sufficiently thin wall.

In conclusion, we studied the domain wall oscillation due to spin-transfer torque by numerically solving the LLG equation. For a domain wall whose rotation angle $\theta_{\max}$ is less than 180°, we found the existence of the critical current density above which the magnetization dynamics are induced. We also found that the critical current density is (i) proportional to $180° - \theta_{\max}$, (ii) approximately proportional to α, and (iii) increased rapidly with decreasing the width of the system.

Acknowledgments

The authors would like to acknowledge the valuable discussions they had with J. Sato, K. Matsushita, N. Yokoshi, H. Tsukahara, H. Arai, and T. Kaneko. This work was supported by JSPS.

References

[1] Franchin M, Fischbacher T, Bordignon G, de Groot P and Fangohr H 2008 *Phys. Rev. B* **78** 054447
[2] Matsushita K, Sato J, Imamura H and Sasaki M 2010 *J. Magn. Soc. Jpn.* **34** 323
[3] Levy P M and Zhang S 1997 *Phys. Rev. Lett.* **79** 5110
[4] Simanek E 2003 *Phys. Rev. B* **63** 224412
[5] Matsushita K, Sato J and Imamura H 2008 *IEEE. Trans. Magn.* **44** 2616
[6] Sato J, Matsushi K and Imamura H 2008 *IEEE. Trans. Magn.* **44** 2608
[7] Matsushita K, Sato J and Imamura H *J. Appl. Phys.* **105** 07
[8] Sato J, Matsushi K and Imamura H *J. Appl. Phys.* **105** 07
[9] Taniguchi T and Imamura H 2010 *Phys. Rev. B* **81** 012405
[10] Taniguchi T and Imamura H 2010 Angle dependence of the magnetoresistance of CCP-CPP-GMR system
[11] Takagishi M, Fuke H N, Hashimoto S, Iwasaki H, Kawasaki S, Shinozaki R and Sahashi M *J. Appl. Phys.* **105** 07
[12] Li Z and Zhang S 2004 *Phys. Rev. B* **70** 024417
[13] Zhang S and Li Z 2004 *Phys. Rev. Lett.* **93** 127204

International Conference on Trends in Spintronics and Nanomagnetism (TSN 2010) IOP Publishing
Journal of Physics: Conference Series **292** (2011) 012008 doi:10.1088/1742-6596/292/1/012008

Modeling thermally activated domain wall dynamics in thin magnetic strips with disorder

L. Laurson[1], **A. Mughal**[1,2], **C. Serpico**[1,3], **G. Durin**[1,4] **and S. Zapperi**[1,5]

[1] ISI Foundation, Torino, Italy
[2] Aberystwyth University, UK
[3] Università di Napoli "Federico II", Italy
[4] INRIM, Torino, Italy
[5] IENI-CNR, Milano, Italy

E-mail: lasse.laurson@gmail.com

Abstract. We study the effect of disorder and temperature on the field-driven dynamics of a transverse domain wall occurring in thin and narrow magnetic strips made of a soft magnetic material such as permalloy. Motivated by a micromagnetic description of such a domain wall, we construct a model based on two coupled flexible lines enclosing the domain wall transition region, capturing both the finite width and the characteristic V-shape of the wall. Disorder is included via randomly distributed pinning centers interacting with the two lines. We study the field-driven dynamics of the domain wall in disordered strips in a finite temperature, and compare our findings to experimental observations of stochastic domain wall dynamics.

1. Introduction

Future forms of memory and logic devices are likely to involve devices such as the racetrack memory [1] and other applications [2, 3, 4] where the local magnetization of narrow and thin magnetic nanostrips or wires is used to store or process information. Such devices rely on the ability to displace the magnetic domain walls in a controlled fashion, by applying field or current pulses. In practice, any material from which such wires and strips are made of will include various sources of disorder, such as roughness of the edges and surfaces of the sample, as well as various point-like impurities present in the bulk of the wire/strip. Such disorder, together with thermal effects, will affect the domain wall dynamics, especially for low driving fields. Thus, understanding and controlling the effect of such imperfections presents a challenge with possible implications also for various practical applications of domain wall dynamics in wires or strips. Previous studies addressing the effect of disorder in nanostrips have focused on the effect of edge roughness [5, 6, 7] and notches [8]. However, the majority of theoretical studies of domain wall dynamics in such systems has been considering "perfect" systems, free of any imperfections which in practice are always present in any realistic sample.

In this paper we present a simplified model of the so-called transverse domain wall occurring in thin and narrow nanostrips made of a soft magnetic material such as permalloy [9], taking into account "bulk" imperfections. The model is based on two flexible lines enclosing the transition region of the domain wall and provides a useful intermediate description between full micromagnetic modeling and the so-called $1d$ models in which the domain wall is reduced into a $1d$ magnetization profile [10]. In particular, the model captures the finite width and the characteristic shape of the domain wall resembling a letter V [11]. Disorder describing e.g. thickness fluctuations of the strip can also be included in a natural way, as randomly distributed pinning centers interacting with the two lines. The paper is organized as follows: In the following Section, the line-based model is presented, while in Section 3, some preliminary numerical results will be presented and compared qualitatively to experimental observations. In Section 4 we present our conclusions.

2. The line-based model

We consider here a strip of length L, width W and thickness D, satifying $L \gg W \gg D$, made of a soft magnetic material such as permalloy. As the anisotropy energy in soft magnetic materials is dominated by shape anisotropy, the domains lie along the long axis of the strip. We focus here on the case where W and D are sufficiently small such that the stable domain wall structure is the so-called transverse wall [9].

Due to the balance between exchange and magnetostatic interactions, the equilibrium shape of the transverse domain wall resembles the letter V. Here we consider a simplified description of such a wall structure by taking the wall to be composed of two coupled flexible lines enclosing the transition region of the domain wall (see Fig. 1). Within this approximation the magnetic charges of the transverse wall are taken to be concentrated along these two lines (with the charge density depending on the local orientation of the line) and at the strip edges. These charges give rise to demagnetizing fields $\vec{H}_{dm}$ acting on the segments of the two lines. Further ingredients of the model include line tension γ_w of the lines, representing the energy cost due to deformations of the wall, as well as a repulsive interaction between the two lines, both originating from the exchange interactions.

2.1. The 1d model

As our model is a generalization of the so-called $1d$ model for a 180° domain wall in a wire, we will first briefly review its main features [10]. The magnetization profile $\vec{m}(x)$ along the wire (x axis) is described by the polar and azimuthal angles θ and ϕ via the ansatz

$$\theta(x,t) = 2\tan^{-1}\exp\left(\frac{x-q(t)}{\Delta(t)}\right), \quad (1)$$

$$\phi(x,t) = \phi(t), \quad (2)$$

where $q(t)$, $\Delta(t)$ and $\phi(t)$ are the domain wall position, width and (out-of-plane) magnetization angle, respectively. By expressing the various micromagnetic energy contributions (such as exchange, anisotropy and applied field) in terms of these *collective coordinates*, one can derive from a Lagrangian formulation the equations of motion, or the *Slonczewski equations*, for $q(t)$, $\phi(t)$ and $\Delta(t)$ [10]:

$$\alpha\frac{\dot{q}}{\Delta} + \dot{\phi} = \gamma_0 H_a, \quad (3)$$

$$\frac{\dot{q}}{\Delta} - \alpha\dot{\phi} = \gamma_0 H_K \frac{\sin 2\phi}{2}, \quad (4)$$

$$\dot{\Delta} = \frac{\gamma_0}{\alpha\mu_0 M_s a}\left[\frac{A}{\Delta} - (K_0 + K\sin^2\phi)\Delta\right]. \quad (5)$$

Here, α is the damping coefficient, γ_0 is the gyromagnetic ratio, H_a is the applied field, K_0 and K are anisotropy constants, H_K is the anisotropy field corresponding to K (i.e. anisotropy perpedicular to the plane), μ_0 is the vacuum permeability, M_s is the saturation magnetization, $a = \pi^2/12$, and A is the exchange constant.

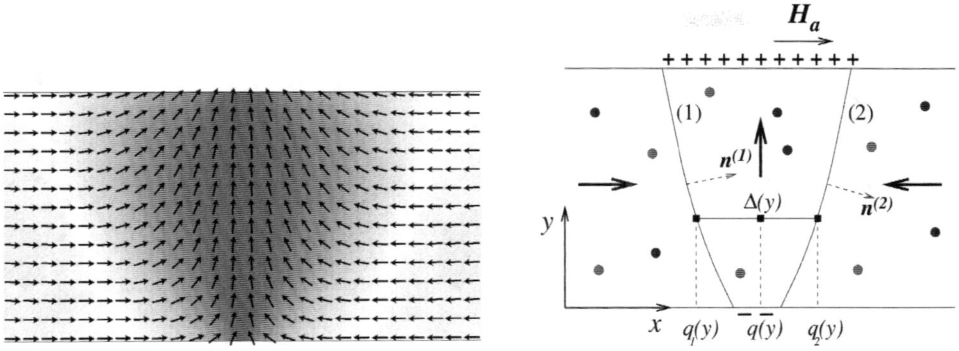

Figure 1. Left: The micromagnetic structure of the transverse domain wall. Right: A schematic of the line-based model, with the two coupled lines (1) and (2) enclosing the domain wall transition region. The model consists of a set of coupled "$1d$ models", with y-dependent widths $\Delta(y)$ and positions $q(y)$, related to the coordinates of the lines $q_1(y)$ and $q_2(y)$ via Eq. (7). Disorder is included via pinning sites which either attract or repel (blue and red dots, respectively) the domain wall.

2.2. Equations of motion for the line-based model

The line-based model is essentially a $2d$ generalization of the $1d$ model. The magnetization is described by the angles $\theta(x, y, t)$ and $\phi(t)$, with the former given by the $2d$ ansatz

$$\theta(x, y, t) = 2\tan^{-1}\exp\left(\frac{x - q(y,t)}{\Delta(y,t)}\right), \tag{6}$$

where the domain wall position $q(y,t)$ and width $\Delta(y,t)$ now depend on the transverse coordinate y. Thus, the domain wall is modeled essentially as a set of $1d$ profiles, with y-dependent positions and widths. The coordinates $q(y,t)$ are related to the coordinates $q_1(y,t)$ and $q_2(y,t)$ of the two lines enclosing the domain wall transition region via

$$q_1(y,t) = q(y,t) - \frac{\Delta(y,t)}{2}, \quad q_2(y,t) = q(y,t) + \frac{\Delta(y,t)}{2}. \tag{7}$$

By employing a similar procedure as in the case of the $1d$ model, one obtains the equations of motion for the coordinates of the two lines,

$$\dot{q}_1 = \frac{\Delta}{\alpha}(\gamma_0 H_a - \dot{\phi}) - \frac{\gamma_0}{2\alpha\mu_0 M_s a}\left[\frac{A}{\Delta} - \Delta K \sin^2\phi\right], \tag{8}$$

$$\dot{q}_2 = \frac{\Delta}{\alpha}(\gamma_0 H_a - \dot{\phi}) + \frac{\gamma_0}{2\alpha\mu_0 M_s a}\left[\frac{A}{\Delta} - \Delta K \sin^2\phi\right], \tag{9}$$

$$\dot{\phi} = \frac{\gamma_0}{1+\alpha^2}H_a - \frac{\gamma_0\alpha}{2(1+\alpha)}H_K \sin 2\phi. \tag{10}$$

In addition, the exchange energy in the y direction leads to an effective line tension which, together with the long-range magnetostatic interactions, couples the different "$1d$ models". The latter are taken into account by approximating the distribution of magnetic charges in the domain wall by concentrating them on the two lines (as if they would represent two 90° domain walls) and along the strip edges, see Fig. 1. These charges then give rise to magnetostatic fields acting on the segments of the two lines similarly to the applied field. Disorder is modeled via randomly distributed pinning centers which exert either an attractive or repulsive force on the

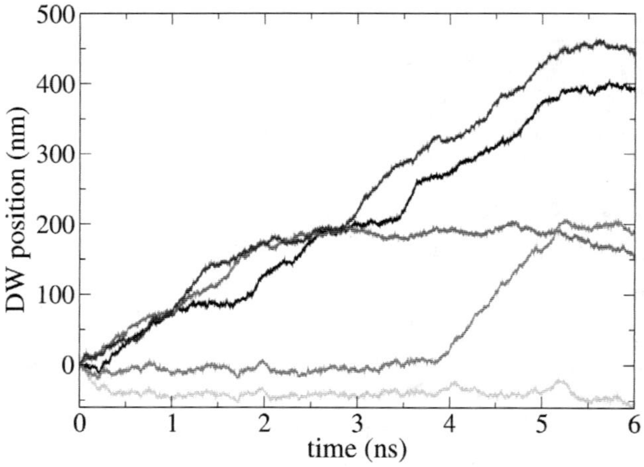

Figure 2. The domain wall position as a function of time due to a 5 ns field pulse starting at $t = 0$. Five different simulation runs with the same parameters but with different realizations of the disorder.

domain wall, representing for instance thickness fluctuations of the strip. In addition, we add a Gaussian white noise term representing the effect of temperature acting on the segments of the two lines. Here we restrict ourselves to consider the field-driven dynamics, but extending the model to include the effect of a spin-polarized electric current is straightforward. Also, only fields well below the Walker field H_W [12] are considered, as the model does not capture the domain wall magnetization reversal taking place for $H_a > H_W$ in the strip geometry via nucleation and propagation of an antivortex [10].

3. Numerical simulations

We investigate the thermally activated creep motion of the transverse wall by integrating numerically the equations of motion (8-10). We use material parameters of a soft magnetic material such as permalloy, i.e. $\alpha = 0.01$, $A = 1.5 \cdot 10^{-11}$ J/m, and $M_s = 8.0 \cdot 10^5$ A/m. To study the effect of disorder and temperature on the domain wall dynamics, we apply a field pulse of 5 ns duration, and monitor the induced displacement of the domain wall. Fig. 2 shows the domain wall position from five simulation runs with different realizations of the disorder with the same parameters. See also the attached video clip of an example simulation run. The domain wall dynamics in this low field regime is clearly stochastic, with a broad distribution of domain wall displacements: sometimes the domain wall does not move at all during the applied field pulse, while some other runs lead to significant displacements of hundreds of nanometers. This is a typical feature of thermally activated subthreshold creep motion: The domain wall spends a significant fraction of the time pinned by the disorder, and occasionally jumps forward due to thermal activation. Such motion is inherently stochastic, and naturally leads to a broad distribution of the domain wall displacements/velocities. Similar stochastic and broadly distributed domain wall velocities have been observed also in experiments [13].

4. Conclusions

Our model serves as a useful intermediate description between full micromagnetic simulations and the $1d$ models, capturing for instance the realistic V-shape of the transverse wall. The present version of the model also has the advantage that realistic physical units and parameters can be used in a transparent way [11]. Also, it is particularly suitable for studies of the effect of disorder on the domain wall dynamics.

Numerical simulations of the model in the low field regime reveal stochastic creep motion of the domain wall, leading to a broad distribution of domain wall displacements/velocities. Such stochastic dynamics is also in qualitative agreement with some recent experimental observations [13]. Future challenges include to understand in more detail the effect of various sources of disorder and thermal fluctuations on the domain wall dynamics in various regimes, including the low field/current creep regime studied here as well as the high driving force regime above the Walker breakdown. Of particular importance for practical applications is also the effect of thermal fluctuations in the case of zero applied fields/currents, which should limit the life time of the "magnetic bits" stored in nanoscale wires and strips.

References

[1] Parkin S S P, Hayashi M and Thomas L 2008 *Science* **320** 190
[2] Beach G S D, Tsoi M and Erskine J L 2008 *J. Magn. Magn. Mater.* **320** 1272
[3] Beach G S D, Nistor C, Knutson C, Tsoi M and Erskine J L 2005 *Nature Mater.* **4** 741; Cowburn R and Petit D 2005 *Nature Mater.* **4** 721
[4] Allwood D A, Xiong G, Faulkner C C, Atkinson D, Petit D and Cowburn R P 2005 *Science* **309** 1688
[5] Martinez E, Lopez-Diaz L, Torres L, Tristan C and Alejos O 2007 *Phys. Rev.* B **75** 174409
[6] Nakatani Y, Thiaville A, and Miltat J 2003 *Nature Mat.* **2** 521
[7] Bryan M T, Atkinson D, and Cowburn R P 2005 *J. Phys: Conf. Ser.* **17** 40
[8] Martinez E, Lopez-Diaz L, Alejos O, Torres L and Tristan C 2007 *Phys. Rev. Lett.* **98** 267202
[9] Nakatani Y, Thiaville A and Miltat J 2005 *J. Magn. Magn. Mater.* **290-291** 750
[10] Thiaville A and Nakatani Y 2006 *Spin dynamics in Confined Magnetic Structures III, Topics Appl. Physics* **101** Ed B Hillebrands and A Thiaville (Berlin Heidelberg: Springer-Verlag) 161-205
[11] Laurson L, Mughal A, Durin G and Zapperi S 2010 *IEEE Trans. Magn.* **46** 262
[12] Schryer N L and Walker L R 1974 *J. Appl. Phys.* **45** 5406
[13] Meier G, Bolte M, Eiselt R, Krüger B, Kim D H and Fischer P 2007 *Phys. Rev. Lett.* **98** 187202

Journal of Physics: Conference Series **292** (2011) 012009 doi:10.1088/1742-6596/292/1/012009

The Effect of a Pulsed Magnetic Field on Domain Wall Resistance in Magnetic Nanowires

R Majidi[1], M M Tehranchi[1,2], A Phirouznia[3], and K Ghafoori Tabrizi[1]

[1] Department of Physics, G.C., Shahid Beheshti University, Evin, 19838-63113, Tehran, Iran
[2] Laser and Plasma Research Institute, G.C., Shahid Beheshti University, Evin, 19839-63113 Tehran, Iran
[3] Department of Physics, Azarbaijan University of Tarbiat Moallem, 53714-161 Tabriz, Iran

E-mail: Teranchi@cc.sbu.ac.ir

Abstract. The effect of a pulsed magnetic field on domain wall magnetoresistance for an ideal one-dimensional magnetic nanowire with a domain wall has been investigated. The analysis has been based on the Boltzmann transport equation, within the relaxation time approximation. The results indicate that the domain wall resistance increase when enhancing the magnetic field. The evaluation of local magnetization has been considered in the presence of a pulsed magnetic field. The time evaluation of the magnetization also has an effect on the domain wall resistance. The resistance depends on the contribution of the Zeeman and exchange interactions.

1. Introduction

Currently there is a great interest in spin dependent transport phenomena in magnetic systems due to the potential application of these phenomena in spin electronic devices [1]. Among magnetic systems, nanowires with their unique transport properties have a special position in spintronics [2]. In these systems, spin dependent scattering plays a significant role in electrical resistance; named magnetoresistance (MR). The MR associated with nanowires containing regions of non-collinearity such as domain walls (DWs) is an outstanding problem [3]. The study of the effects of different scattering mechanisms on resistance and controlling the DWR has gained crucial importance [4-8]. For instance, the DWR of a ferromagnetic nanowire has been investigated in the presence of an external magnetic field [5, 7].

For practical applications, DWR when affected by magnetic fields should lead to new insights into its control. However, DWR has only been investigated in constant external magnetic fields and the effects of pulsed magnetic fields on DWR are still unclear. In this paper, we have studied the contribution of DW to the resistance of magnetic nanowires in the presence of a pulsed magnetic field. The analysis has been based on the Boltzmann transport equation, within the relaxation time approximation.

2. Theoretical Considerations

2.1. Description of Model and Interactions

We have studied a 180° Néel-type DW between two ferromagnetic regions with opposite directions of magnetization (Figure 1). For this ideal one-dimensional DW, the rotation angle can be written as $\theta(z) = (\pi / d)z$, which describes the angle between the local direction of the magnetization, $\hat{M}_0(\vec{r})$, and the wire direction, z-axis.

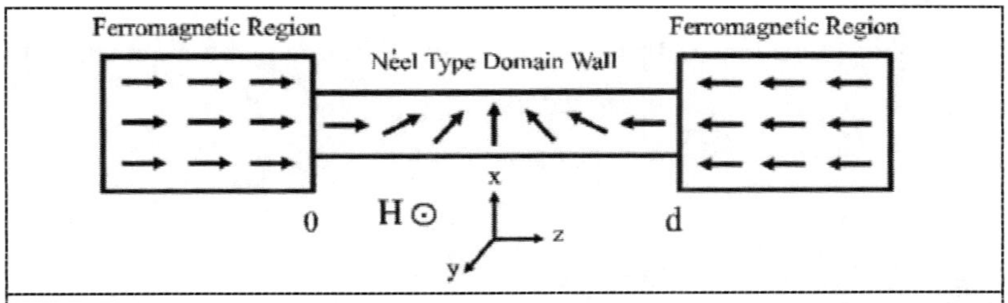

Figure 1. A 180° Néel-type DW between two ferromagnetic regions.

The Hamiltonian of the system in the presence of an external magnetic field can be written as:

$$H = H_0 + H_{ex} + H_{\mathrm{H}}. \tag{1}$$

The first term H_0 contains kinetic energy and nonmagnetic periodic potential, $V(z)$, as follows:

$$H_0 = -\frac{\hbar^2}{2m}\frac{d^2}{dz^2} + V(z). \tag{2}$$

The second term H_{ex} represents the exchange interaction between the conduction electrons and the localized magnetic moments, and can be expressed as:

$$H_{ex} = -\Delta_{ex}\hat{\sigma}.\hat{M}(\vec{r},t), \tag{3}$$

in which Δ_{ex} is the exchange interaction strength and $\hat{\sigma}$ denotes the spin operators in terms of the Pauli spin matrices. We considered a system which has the equilibrium value $\hat{M}_0(\vec{r})$, and asked what the response would be to a time varying transverse magnetic field $\mathrm{H}(t)$. We write magnetization as $\hat{M}(\vec{r},t) = \hat{M}_0(\vec{r}) - \hat{m}(\vec{r},t)$ in the presence of a pulsed magnetic field. The unit vector along the direction of the local magnetization is introduced by $\hat{M}(\vec{r},t)$ in the presence of a pulsed magnetic field. Therefore, the term of the exchange interaction is:

$$\begin{aligned} H_{ex} &= H_{ex}(0) + H_{ex}(\hat{m}) \\ &= -\Delta\hat{\sigma}.[\hat{M}_0(\vec{r}) - \hat{m}(\vec{r},t)]. \end{aligned} \tag{4}$$

In the presence of a pulsed magnetic field the time evaluation of the local magnetization is:

$$\frac{d\hat{M}(\vec{r},t)}{dt} = -\gamma\,\hat{M}(\vec{r},t)\times\mathrm{H}(t) + \alpha\,\hat{M}(\vec{r},t)\times\frac{d\hat{M}(\vec{r},t)}{dt}. \tag{5}$$

When the field is turned off the local magnetization is considered as:

$$\hat{M}(\vec{r},t) = \hat{M}_0(\vec{r}) - \hat{m}(\vec{r},\Gamma)\,e^{-(t-\Gamma)/T}, \tag{6}$$

where $\gamma = g\mu_B / \hbar$ is the gyromagnetic ratio and α is a phenomenological damping constant. For simplicity, the equation (5) is linearized by neglecting terms which are quadratic in $\mathrm{H}(t)$ or the components of $\hat{M}(\vec{r},t)$.

The last term in the Hamiltonian is the Zeeman interaction,

$$H_{\mathrm{H}} = -\Delta_{\mathrm{H}}\hat{\boldsymbol{\sigma}}.\hat{n}_{\mathrm{H}}, \tag{7}$$

where $\Delta_{\mathrm{H}} = g\mu_B \mathrm{H}(t)/2$, μ_B is the Bohr magneton and $\hat{n}_{\mathrm{H}}$ is the unit vector along the external magnetic field. The rectangular pulsed magnetic field with an amplitude of H_0 and duration of Γ is considered as:

$$\mathrm{H}(t) = \begin{cases} \mathrm{H}_0 & 0 \le t \le \Gamma \\ 0 & t < 0, \Gamma > t \end{cases}. \tag{8}$$

Using an approach based on the perturbation method of Ref. [4], the eigenstates of $H_0 + H_{ex}(0)$ for a one-dimensional system may be formulated as:

$$\left|\psi_k^{\uparrow}\right\rangle = \frac{\tilde{\alpha}(k)}{\sqrt{d}} e^{ikz} R_{\theta(z)} \begin{pmatrix} 1 \\ ik\zeta \end{pmatrix}, \tag{9a}$$

$$\left|\psi_k^{\downarrow}\right\rangle = \frac{\tilde{\alpha}(k)}{\sqrt{d}} e^{ikz} R_{\theta(z)} \begin{pmatrix} ik\zeta \\ 1 \end{pmatrix}, \tag{9b}$$

where $\tilde{\alpha}(k) = (1 + k^2\zeta^2)^{-1/2}$ and $\zeta = \pi^2\hbar^2/(8m\Delta d^2)$ are the normalization coefficient and the perturbation parameter, respectively. Here, $R_{\theta(z)} = \exp(-i\theta(z)\hat{\sigma}.\hat{n}_{\mathrm{H}}/2)$ is the spin rotation operator about y-axis corresponding to Figure 1.

2.2. Spin dependent transport characteristics

To calculate the DWR, we consider the Boltzmann equation in the relaxation time approximation for finding the deviation from equilibrium distribution function.

Using the approach based on the method used in Ref. [8], the spin dependent relaxation time will be simplified by:

$$\tau^{\uparrow}(k) = \frac{\hbar^3}{md} \frac{kk_-\left(k\left|V_{k_+,k}^{\uparrow\downarrow}\right|^2 + k_+\left|V_{k_-,k}^{\downarrow\uparrow}\right|^2\right)}{\left|V_{k_-,k}^{\downarrow\uparrow}\right|^2 \cdot \left|V_{k_+,k}^{\uparrow\downarrow}\right|^2 \left(k^2 - k_+k_-\right)}, \tag{10a}$$

$$\tau^{\downarrow}(k) = \frac{\hbar^3}{md} \frac{kk_+\left(k_-\left|V_{k_+,k}^{\uparrow\downarrow}\right|^2 + k\left|V_{k_-,k}^{\downarrow\uparrow}\right|^2\right)}{\left|V_{k_-,k}^{\downarrow\uparrow}\right|^2 \cdot \left|V_{k_+,k}^{\uparrow\downarrow}\right|^2 \left(k^2 - k_+k_-\right)}, \tag{10b}$$

in which $k_{\pm} = \sqrt{k^2 \mp 4m\Delta/\hbar^2}$. The value of $\sqrt{4m\Delta/\hbar^2}$ is small enough in relation to the Fermi wave vector, k_F.

The matrix elements of scattering potentials, $V_{k'k}^{\sigma'\sigma}$, can be determined by considering the Zeeman term as scattering potential,

$$V_{k',k}^{\sigma',\sigma} = \left\langle \psi_{k'}^{\sigma'} \middle| H_{\mathrm{H}} + H_{ex}(\hat{m}) \middle| \psi_k^{\sigma} \right\rangle \qquad (0 < t < \Gamma) \tag{11a}$$

$$V_{k',k}^{\sigma',\sigma} = \tilde{\alpha}'\tilde{\alpha} \frac{\exp[(k-k')d] - 1}{i(k-k')d} \left(-\Delta_{\mathrm{H}} + \Delta_{ex}\alpha\gamma\,\mathrm{H}_0 t\right) \begin{pmatrix} (k'+k)\zeta & -i(1-k'k\zeta^2) \\ i(1-k'k\zeta^2) & -(k'+k)\zeta \end{pmatrix}, \tag{11b}$$

and,

$$V_{k',k}^{\sigma',\sigma} = \left\langle \psi_{k'}^{\sigma'} \middle| H_{ex}(\hat{m}) \middle| \psi_k^{\sigma} \right\rangle \qquad (t > \Gamma) \tag{12a}$$

$$V_{k',k}^{\sigma',\sigma} = \tilde{\alpha}'\tilde{\alpha}\frac{\exp[(k-k')d]-1}{i(k-k')d}\Delta_{ex}\alpha\gamma\,\mathrm{H}_0\Gamma\begin{pmatrix}(k'+k)\zeta & -i(1-k'k\zeta^2)\\ i(1-k'k\zeta^2) & -(k'+k)\zeta\end{pmatrix}e^{-(t-\Gamma)/T}, \tag{12b}$$

here, T is the spin lattice relaxation time.

Using $\tau^{\sigma}(k)$, the resistivity per unit length can be determined as follows:

$$\Re = \left[\frac{e^2\hbar^2}{2\pi m^2}\sum_{\sigma}\int dk\, k^2\tau^{\sigma}(k)\delta(\varepsilon_{k\sigma}-\varepsilon_F)\right]^{-1}, \tag{13}$$

where ε_F is the Fermi energy.

3. Results and Discussion

In our calculations, the parameters had been chosen in such a way that the condition for validity of the semiclassical approximation ($k_F d >> 1$) would be applicable. We have used the typical acceptable parameters: $d = 20\,nm$, $\varepsilon_F = 10\,eV$, $\Delta_{ex} = 0.1\,eV$, and $m = m_e$.

The resistance per unit length of DW versus time in the presence of the pulsed magnetic field is shown in Figure 2. Both increasing and decreasing of the DWR have been observed depending on the contribution of the Zeeman and exchange interactions. When the magnetic field is applied, the change in the transverse component of the magnetization, M_y, is small, so the effect of the Zeeman interaction is more significant than that of the exchange interaction. It should be note that increasing M_y enhances the width of DW, and this enhancement results in the reduction of DWR. A similar trend is observed in previous studies [6, 7] which indicates DWR reduction when increasing the width of DW. After some time, M_y becomes significant and the exchange interaction plays an important role in DWR. In this case, DWR is increased by enhancing M_y.

The resistance per unit length of DW versus time in the absence of a magnetic field is presented in Figure 3. When the pulse is turned off, the magnetization returns to its equilibrium value of $\hat{M}_0$. Therefore, the effect of exchange interaction is decreased and reduction of DWR is observed for $\Gamma < t$.

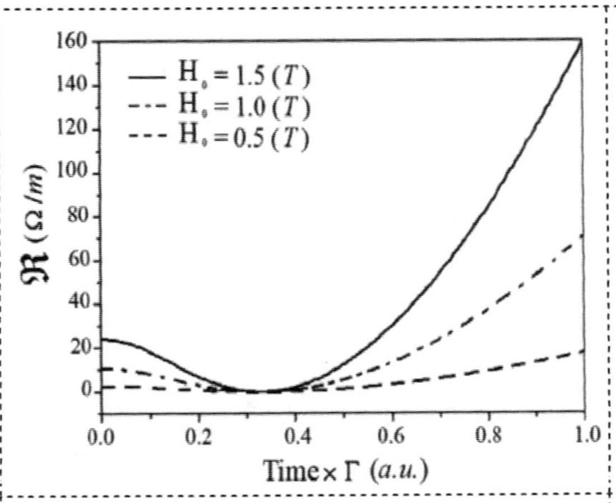

Figure 2. The resistance per unit length of DW versus time for different values of the magnetic field. It was calculated for $d = 20\,nm$ and $0 \le t \le \Gamma$.

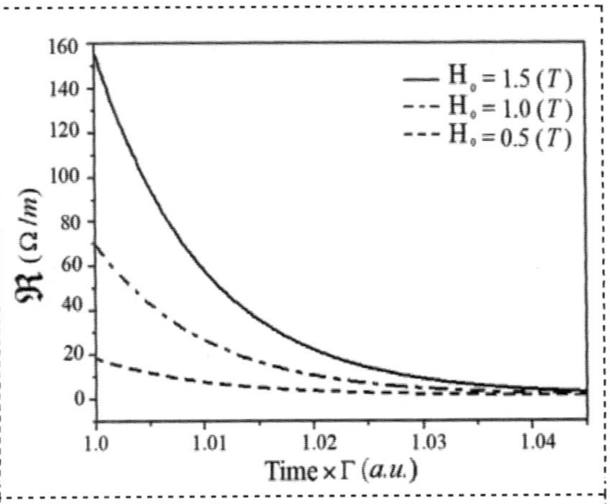

Figure 3. The resistance per unit length of DW versus time for different values of the magnetic field. It was calculated for $d = 20\,nm$ and $\Gamma < t$.

As can be seen in Figures 2 and 3, the external magnetic field parallel to the y-axis significantly affects DWR. The increase of the resistance resulting from DW has been observed with the enhancement of the magnetic field.

4. Conclusions

We have studied DWR in a ferromagnetic nanowire using the Boltzmann equation, within the relaxation time approximation. The one-dimensional Néel-type DW between two ferromagnetic regions with antiparallel magnetization has been considered. We have investigated how the Zeeman and exchange interactions affect the resistance associated to DW. Calculations show that the resistance of DW increases as the magnetic field becomes stronger. In the presence of the pulsed magnetic field, both increasing and decreasing of DWR are observed depending on the contribution of the Zeeman and exchange interactions. When the magnetic field is turned off, DWR shows a fast decrement.

[1] Schmidt G, Gould C, and Molenkamp L W 2004 *Physica* E **25** 150
[2] Sapra S, Sarma D D, Sanvito S, and Hill N A 2002 *Nano Lett.* **2** 605
[3] Wosinski T, Figieski T, Morawski A, Makosa A, Osinniy V, Wrobel J, and Sadowski J 2008 *J. Mater. Sci: Matter. Electron* **19** 111
[4] Levy P M, and Zhang S 1997 *Phys. Rev. Lett.* **79** 5110
[5] Phirouznia A, Tehranchi M M, and Ghanaatshoar M 2006 *Eur. Phys. J.* B **54** 103
[6] Phirouznia A, Tehranchi M M, and Ghanaatshoar M 2007 *Phys. Rev.* B **75** 224403
[7] Ghanaatshoar M, Fallahi V, Tehranchi M M, and Phirouznia A 2008 *IEEE Tran. On Mag.* **44** 3127
[8] Majidi R, Tehranchi M M, Phirouznia A, and Ghafoori Tabrizi K 2010 *Eur. Phys. J.* B **76** 475

Chemical and electronic properties of Fe/MgO/Ge heterostructures for spin electronics

D Petti, M Cantoni, C Rinaldi and R Bertacco

LNESS, Dipartimento di Fisica – Politecnico di Milano, Via Anzani 42, 22100 Como, Italy

matteo.cantoni@polimi.it

Abstract. We report on the chemical and electronic properties of epitaxial Fe/MgO/Ge(001) heterostructures probed by X-ray Photoemission Spectroscopy. At variance with the Fe/MgO/Fe system, annealing at 570 K produces a sizable interdiffusion at the upper Fe/MgO interface, while at 470 K this process is inhibited. The XPS analysis of band alignment in heterostructures annealed at 470 K grown onto an intrinsic Ge substrate indicates that the Fermi level is placed at the center of the MgO gap and that the Schottky barrier height is 0.35±0.1 eV, thus indicating a partial depinning of the Fermi level.

1. Introduction

The injection of a spin polarized current from a ferromagnetic electrode to a non-magnetic semiconductor, or vice versa, requires the presence of an interfacial spin dependent resistance to overcome the problem of the resistivity mismatch [1]. MgO tunneling barriers have been employed to this scope, with performances remarkably better than those obtained by amorphous barriers (Al_2O_3) [2]. For this kind of applications GaAs was initially used, exploiting spin optical pumping, and the epitaxial growth of MgO on GaAs has already been demonstrated [3]. Silicon has been also employed, due to its primary role in electronics and long spin diffusion length, but its relevant chemical reactivity makes the realization of epitaxial structures quite difficult [4]. Germanium, instead, only very recently has deserved some attention [5] though its high mobility, the opportunity of spin manipulation in Ge-based heterostructures [6] and the possibility of spin optical pumping [7] make it a good candidate for the development of semiconductor spintronics. A Ge-based system really promising for spintronics applications is the Fe/MgO/Ge heterostructure, exploiting the very efficient spin filtering at the MgO/Fe interface [8]. In such applications, the quality of the interfaces is fundamental, because disordered and/or smooth interfaces would lead to current depolarisation: a fully epitaxial Fe/MgO/Ge heterostructure with sharp interfaces is then highly desirable.

In this paper we report on the characterization of the chemical and electronic properties of Fe/MgO/Ge(001) epitaxial heterostructures grown by molecular beam deposition. As a matter of fact, apart from the recent work by Zhou et al. [9], the Fe/MgO/Ge structure has not been widely investigated and a detailed characterization of the interfacial chemistry in connection to the deposition condition, as well as a full investigation of the band alignment, is still lacking.

2. Experimental

The sample preparation and characterization have been carried out in-situ in an ultrahigh vacuum (UHV) system equipped with a Molecular Beam Epitaxy (MBE) apparatus and a chamber for electron spectroscopies [10]. Standard n-type Ge(001) ($n=10^{15}$ cm^{-3}) wafers were used as substrates. After ultrasonic washing in isopropanol, the substrates were inserted into the system and annealed at 920 K for 1 hour, with pressure below $2x10^{-9}$ mbar, in order to remove the native oxide. A further annealing at 1000 K for 20 minutes produced an ordered (2x1) Ge(001) surface, as checked by Reflection High Energy Electron Diffraction (RHEED), Low Energy Electron Diffraction (LEED) and X-ray Photoelectron Diffraction (XPD). X-ray Photoemission Spectroscopy (XPS) revealed a negligible residual surface contamination (essentially O and C, below 3% in the probing depth of XPS).
MgO was deposited by MBE at room temperature (RT), at a rate of ~ 0.6 Å/min, by evaporation of pieces of stoichiometric MgO contained in a Ta crucible, heated by electron bombardment. A post-annealing treatment at 770 K for 20 minutes was finally performed in order to improve the MgO crystallinity, as indicated by RHEED and XPD.
The Fe deposition was performed by MBE at RT, at a rate of ~ 2 Å/min, and two post-annealing treatments at 470 K and 570 K where checked, in order to optimize the annealing temperature giving the best crystal quality of the Fe/MgO interface compatible with limited interdiffusion. As in the case of MgO growth, the deposition rate was measured by a thickness monitor, then the film thickness was checked by XPS. During deposition, the pressure in the chamber was in the 10^{-10} Torr range. The Fe thickness in heterostructures fabricated for transport measurements was ~10 nm. Tunnelling junctions with area ranging from 30 to 900 μm^2 have been microfabricated via standard optical lithography and electrical measurements performed in the four probe configuration.

3. Results

Following the recipe described above, a high quality epitaxial MgO barrier has been grown on Ge(001), with a rotation of the square lattice of MgO by 45° with respect of the cubic cell of Ge in order to reduce the lattice mismatch to 5.4% [11]. The optimization of the growth conditions of the top Fe layer has been carried out essentially looking for the optimum annealing temperature after RT deposition of Fe on a MgO/Ge(001) epitaxial structure, previously annealed at 770 K. A 2 nm Fe film was initially grown on a MgO(3 nm)/Ge(001) sample at RT and afterwards two post-annealing treatment were performed (20' at 470 K and 20' at 570 K). As a matter of fact, RT growth followed by a post annealing at 570 K is the procedure used for the deposition of the top electrode in Fe/MgO/Fe heterostructures for magnetic tunnelling junctions [12]; in particular, the scope of the annealing treatment was to reduce the high dislocation density in the upper MgO/Fe interface arising from RT growth of the top Fe layer. For Fe/MgO/Ge(001), however, the second interface is not so stable as in the case of Fe/MgO/Fe because the annealing at 570 K provokes a great modification of the crystalline structure of the upper layers. In table 1 the intensity ratio taken at 0° collection angle (i.e. the angle between the sample normal and the direction of photoelectrons collected by the analyzer) of Fe2p and Mg1s, Fe2p and O1s, Fe2p and Ge3d, O1s and Mg1s XPS peaks after the post-growth annealing are presented.

Table 1. Intensity ratios of XPS intensities from the Ge, MgO and Fe layers for two different annealing temperature (T_a) after RT growth of the Fe layer.

T_a (K)	Fe2*p*/Mg1*s*	Fe2*p*/O1*s*	Fe2*p*/Ge3*d*	O1*s*/Mg1*s*
470	1.17	1.19	1.19	0.98
570	0.34	0.99	1.18	0.34

The ratios are normalized to the corresponding values before annealing. After annealing at 570 K, the Fe2p/Mg1s ratio decreases dramatically to ~ 30% of the initial value, thus indicating a strong intermixing between the Fe and MgO layers. Oddly, the same does not happen for the O1s/Fe2p ratio that remains unchanged (0.99 with respect to the initial ratio at 0° collection angle), nor for the Fe2p/Ge3d ratio which, instead, slightly increases (1.18). The picture arising as a whole is that at 570 K Mg atoms diffuse into the Fe layer, and in fact also the O1s/Mg1s ratio is strongly altered, as clearly visible in the last column of table 1.

The annealing at 470 K, on the contrary, preserves the chemical sharpness of the interface, with the ratios between the chemical concentration seen by XPS staying essentially constant. The slight increase of the Fe2p/Mg2s, Fe2p/O1s and Fe2p/Ge3d (on the order of 15-20%) is compatible with a rearrangement of the Fe layer, corresponding to a better wetting of the MgO surface induced by annealing. The limited interaction between Fe and MgO at 470 K is confirmed by a detailed analysis of the line shape of Fe2p peaks. For these measurements the Mg-kα line (1253.6 eV) was used and the analyzer set in order to have a total energy resolution of 0.84 eV. Contributions due to α_3 and α_4 satellites of the Mg-Kα line were subtracted from the original spectrum. Figure 1 shows Fe2p spectra, taken at 0° collection angle in order to maximize the sensitivity to the buried Fe/MgO interface, from the sample after MgO deposition (a) and after annealing at 470 K (b). Data (black dots) have been fitted with the lineshape of a Fe2p peak taken on a clean Fe(001) single crystal in the same experimental condition (continuous red curve with the $Fe2p_{3/2}$ peak located at the nominal position of metallic iron, i.e. 706.7 eV), in order to evidence additional contributions to the spectra typically arising from oxidized interfacial Fe atoms. Unfortunately, the presence of O KLL peak, partially superposed to Fe2p tail at high binding energy, prevents from a precise background subtraction and accurate deconvolution of the spectra with peaks representative of different Fe oxidation states. However, from a rough analysis it is clear that at least two additional contributions are present in all the spectra: Fe^{2+} and Fe^{3+} whose energy positions are 709.7 eV and 711.1 eV respectively. A rough quantitative estimation of the interfacial oxidation can be done looking at the ratio of the shaded area, i.e. the difference between the data and the fit with a nominal spectrum from clean Fe, to the Fe2p intensity after background subtraction. This ratio is on the order of 10±1% before and after annealing at 470 K, thus confirming that the latter does not produce additional interfacial oxidation.

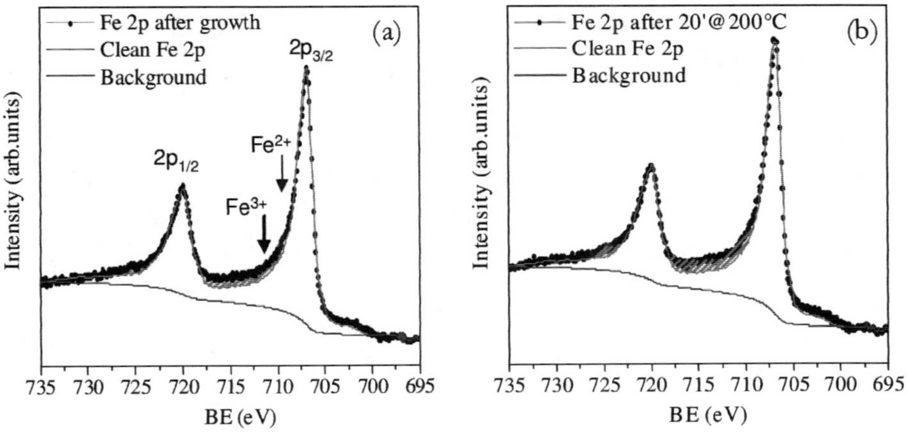

Figure 1: Fe2p XPS spectra taken with Mg-ka radiation on a Fe/MgO/Ge heterostructure immediately after RT growth of the Fe layer (a) and after annealing at 470 K (b).

The determination of the band alignment has been carried out looking at the energy position of core levels in the Ge, MgO and Fe layers by XPS. The method consists in the evaluation of the valence band offset (VBO) at the different interfaces, using a core-level photoemission-based method, already used for the Fe/MgO/GaAs heterostructure by Lu et al. [13]. As depicted in Figure 2(a), at the interface between MgO and Ge the VBO can be obtained as:

$$\Delta E_V = \left(E_{Ge3d}^{Ge} - E_{VBM}^{Ge}\right) - \left(E_{O2s}^{MgO} - E_{VBM}^{MgO}\right) - \Delta E_{CL} \tag{1}$$

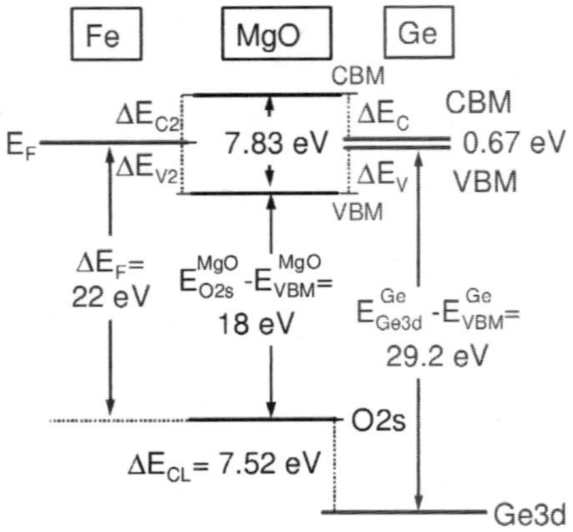

Figure 2(a). Band offsets at the interfaces Fe/MgO and MgO/Ge.

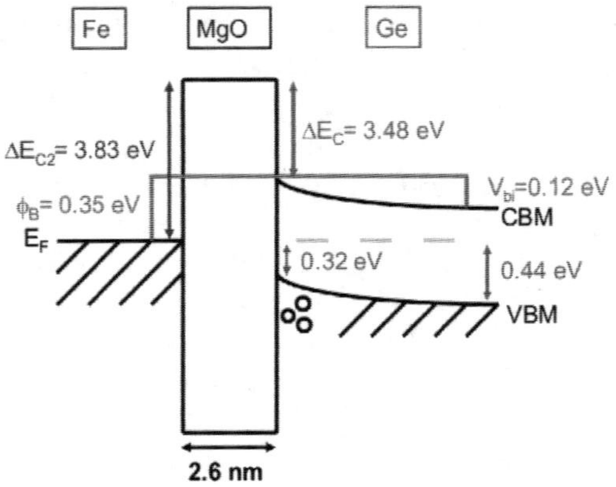

Figure 2(b). Band diagram in the Fe/MgO/Ge heterostructure

$\Delta E_{CL} = E^{Ge}_{Ge3d} - E^{MgO}_{O2s}$ is the energy difference between the Ge3d and O2s core levels measured in the heterojunction, while $(E^{MgO}_{O2s} - E^{MgO}_{VBM})$ and $(E^{Ge}_{Ge3d} - E^{Ge}_{VBM})$ are the energy positions of the respective peaks in Ge and MgO bulk with respect to the valence band maxima (VBM). For $(E^{Ge}_{Ge3d} - E^{Ge}_{VBM})$ we employed the tabulated values of the core level energy position, while for $(E^{MgO}_{O2s} - E^{MgO}_{VBM})$ we measured the valence band XPS spectrum from a MgO single-crystal sample and we found the VBM by employing the Kraut method [14], which involves the fit of the valence band spectrum with a Gaussian broadened DOS in this energy region [13]. All the heterojunction core level peaks have been fitted with the lineshape of the respective core levels in the bulk, in order to achieve a better precision in the energy position. Finally, the difference $\Delta E_F = E^{MgO}_{O2s} - E_F$ has been estimated from the XPS spectrum taken on the entire heterostructure as the energy difference between the Fermi edge and the O2s peak. For the evaluation of the conduction band offset (CBO), we employed 7.83 eV for the band gap of MgO [13] and 0.67 eV for that of Ge.

The VBOs at the MgO/Ge and Fe/MgO interfaces result, respectively, $\Delta E_V = 3.68$ eV and $\Delta E_{V2} = 4$ eV, while the corresponding CBOs are $\Delta E_C = 3.48$ eV and $\Delta E_{C2} = 3.83$ eV. Figure 2(b) shows the band diagram obtained from XPS data, assumed as representative of band alignment at the interface due to the limited probe depth of XPS, and from the calculated value of the bulk VBO in Ge (0.44 eV) for the n-type Ge(001) substrate, with $n=10^{15}$ cm^{-3}, employed in this work [15]. Note that the Fermi level is placed in the middle of the MgO gap, so that there is an efficient barrier for tunnelling of electrons and holes. Furthermore a clear upwards band bending is seen at the MgO/Ge interface which results in the formation of a built-in voltage V_{bi}=0.12 eV and in a Schottky barrier height (SBH) of 0.35±0.1 eV. The latter value of the SBH is lower than that observed in metal-Ge Schottky junctions where the strong pinning of the Fermi level at the VBM produces a SBH in the order of 0.6 eV, i.e. roughly coinciding with the Ge gap. The insertion of the MgO layer is then effective in producing a partial depinning of the Fermi level, as already reported in Zhou et al., due to removal of interfacial trapping states. More important, the MgO barrier causes a substantial deviation from diffusive transport, as evidenced by the I(V) and conductance curves reported in Figure 3 from a Fe/MgO/Ge tunnelling junction with area of 600 μm^2. The parabolic conductance is a clear signature of tunnelling which constitutes the key element for the application of Fe/MgO/Ge heterostructures to spin injection/detection in semiconductors.

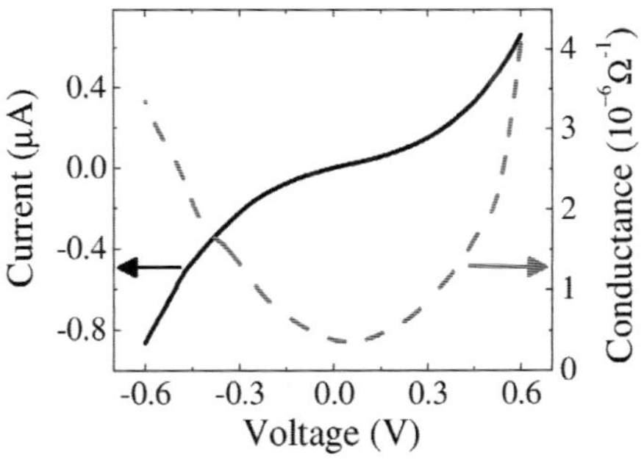

Figure 3. Current-voltage (black continuous line) and conductance-voltage (red dashed line) characteristics of a 600 μm^2 Fe/MgO/Ge junction, measured at room temperature.

4. Conclusions

By means of a careful XPS analysis we found that the chemical sharpness of the upper Fe/MgO interface is preserved for annealing at temperature not exceeding 470 K, at variance with the Fe/MgO/Fe system where the optimum annealing temperature producing high tunneling magnetoresistance while preserving sharp interfaces is 570 K. The investigation by XPS of the band alignment in the Fe/MgO/Ge heterostructure in case of slightly n-doped Ge shows that the Fermi level is placed in the middle of the MgO gap, while the Schottky barrier height (SBH) results 0.35±0.1 eV. The band alignment is then suitable for tunnelling transport (both for holes and electrons), while the SBH lowering indicates a partial depinning of the Fermi level in Ge due to the insertion of the MgO layer. These conditions are particularly favourable to spin dependent transport across the junction, as confirmed by transport measurements on microfabricated tunnelling junctions which essentially display a tunnelling behaviour.

5. Acknowledgments

This work was funded by Fondazione Cariplo via the project MANDIS (Project No. 2007.5095).

[1] Fert A and Jaffrès H 2001 *Phys. Rev. B* **64** 184420
[2] Jiang X, Wang R, Shelby R M, Macfarlane R M, Bank S R, Harris J S and Parkin S S 2005 *Phys. Rev. Lett.* **94** 056601
[3] Hung L S, Zheng L R and Blanton T N 1992 *Appl. Phys. Lett.* **60** 3129
[4] Miao G X, Chang J Y, van Veenhuizen M J, Thiel K, Seibt M, Eilers G, Münzenberg M and Moodera J S 2008 *Appl. Phys. Lett.* **93** 142511
[5] Patibandla S, Atkinson G M, Bandyopadhyay S and Tepper G C 2010 *Physica E* **42** 1721
[6] Ganichev S D, Danilov S N, Bel'kov V V, Giglberger S, Tarasenko S A, Ivchenko E L, Weiss D, Jantsch W, Schäffler F, Gruber D and Prettl W 2007 *Phys. Rev. B* **75** 155317
[7] Loren E J, Ruzicka B A, Werake L K, Zhao H, van Driel H M and Smirl A L 2009 *Appl. Phys. Lett.* **95** 092107
[8] Butler W H, Zhang X-G, Schulthess T C and MacLaren J M 2001 *Phys. Rev. B* **63** 054416
[9] Zhou Y, Ogawa M, Bao M, Han W, Kawakami R K and Wang K L 2009 *Appl. Phys. Lett.* **94** 242104
[10] Bertacco R, Cantoni M, Riva M, Tagliaferri A and Ciccacci F 2005 *Appl. Surface Science* **252** 1754
[11] Cantoni M, Petti D, Rinaldi C and Bertacco R, submitted to *Microelectronic Engineering*
[12] Yuasa S, Nagahama T, Fukushima A, Suzuki Y and Ando K 2004 *Nature. Mater.* **3** 868
[13] Lu Y, Breton J C L, Turban P, Lepine B, Schieffer P and Jezequel G 2006 *Appl. Phys. Lett.* **88** 042108; 2006 *Appl. Phys. Lett.* **89** 152106
[14] Kraut E A, Grant R W, Waldrop J R and Kowalczyk S P 1983 *Phys. Rev. B* **28** 1965
[15] Sze S M and Ng K K 2007 *Physics of Semiconductor Devices 3rd ed.*(Wiley–Interscience)

Journal of Physics: Conference Series **292** (2011) 012011 doi:10.1088/1742-6596/292/1/012011

Magnetic anisotropy of (Ge,Mn) nanostructures

A. Jain, M. Jamet, A. Barski, T. Devillers, I.-S. Yu, C. Porret, S. Gambarelli, V. Maurel, G. Desfonds, J.F. Jacquot

Institut Nanosciences et Cryogénie, CEA-UJF, F-38054, Grenoble, France

E-mail: abhinav.jain@cea.fr

Abstract. We present a correlation between structural and magnetic properties of different (Ge,Mn) nanostructures grown on Ge(001) and Ge(111) substrates. Thin films of $Ge_{1-x}Mn_x$ were grown by low temperature molecular beam epitaxy to favor the out-of-equilibrium growth. Depending on the growth conditions crystalline or amorphous (Ge,Mn) nanocolumns have been observed on Ge(001) substrates. The magnetic properties were probed by superconducting quantum interference device (SQUID), vibrating sample magnetometer (VSM) and electron paramagnetic resonance (EPR). With the help of these complementary techniques (SQUID and EPR), magnetic anisotropy in these nanostructures has been investigated and different anisotropy constants were calculated.

1. Introduction

The electric control of magnetic properties in Ferromagnetic Semiconductors (FMS) make them very interesting materials for their potential use in spintronics industry [1, 2]. Making a FMS with high Curie temperature (T_c) and compatible with mainstream silicon technology is highly desirable. Up to now major efforts have been focused on diluted magnetic semiconductors (DMSs) in which the host semiconducting matrix is randomly substituted by transition metal, the model system being Mn-doped III-V semiconductor GaAs [3]. However, T_c in these DMSs remain rather low and increasing the transition metal concentrations further to increase the Tc leads to precipitation of secondary phases. Moreover, the understanding of origin of hole-mediated ferromagnetism in (III,Mn)V DMS is still under debate.

Recently group IV semiconductor Ge doped with Mn is gaining attention as a FMS due to its high Curie temperature, compatibility with existing Si technology and reported to be half metallic from the band structure calculations [4, 5]. The first evidence of ferromagnetism in (Ge,Mn) was reported by Park et. al. where the highest Tc was reported to be 116K [6]. Since then (Ge,Mn) system has been studied intensively and the solubility of Mn in Germanium has been enhanced by non-equilibrium growth techniques [7, 8, 9, 10, 11, 12, 13, 14, 15].

In our case we have Mn-rich nanocolumns in $Ge_{1-x}Mn_x$ films grown by low temperature molecular beam epitaxy (LT-MBE) on Ge(001) substrates [4, 11]. These self-organized nanocolumns exhibit high T_c and are formed by 2D spinodal decomposition. This columnar growth forming transition metal rich nanostructures exhibiting high T_c has been predicted by ab-initio calculations coupled with kinetic Monte-Carlo simulations [16, 17]. Also Ge_3Mn_5 thin-films on Ge(111) have been grown. In this paper, we discuss the structural and magnetic properties of these (Ge,Mn) nanostructures. We investigate the magnetic anisotropy and extract various anisotropy constants.

2. Sample growth and experimental details

The (Ge,Mn) films were grown by MBE on epi-ready Ge(001) substrates at low growth temperature T_g to stabilize metastable nanocolumns which are either crystalline (T_g=100°C) or amorphous (T_g=150°C). The growth process is described in detail in Ref. [11]. Also epitaxial Ge_3Mn_5 films were grown at 250°C on Ge(111) substrates by co-depositing Ge and Mn atoms according to the Ge_3Mn_5 stoechiometry.

Superconducting quantum interference device (SQUID), vibrating sample magnetometer (VSM) and electron paramagnetic resonance (EPR) techniques were used to calculate the magnetic anisotropy constants. EPR is a very sensitive spectroscopy technique with high field resolution, where the magnetization dynamics is described by the Landau-Lifschitz-Gilbert equation [18]. However, it is much convenient to use Smit Beljers formalism [19, 20], where the equation of motion is expressed as a function of the total free energy density F (equations are given in ref. [21, 22]). These equations are used to calculate the magnetic anisotropy fields. However, SQUID measurements are needed to obtain the saturation magnetization and hence calculate the anisotropy constants. This approach has proved very useful in studying magnetic anisotropy in metallic thin-films and (III,Mn)V DMS [22, 21]. However, in our case we have (Ge,Mn) nanostructures but still the same approach has been used. We have tested our method to determine magnetic anisotropy constants using both EPR and SQUID measurements on Ge_3Mn_5 films grown on Ge(111) substrates [23].

The EPR experiments were performed at T = 5 K using a Bruker ESP 300 spectrometer. The spectrometer was operating at X-band (9.4 GHz) and Q-band (34 GHz), with a modulation frequency of 100 kHz and a modulation amplitude of 10 Oe for lock-in detection. The measurements performed with the out-of-plane applied field swept from [001] to [1$\bar{1}$0] crystal direction. In this paper, we denote the effective out-of-plane second-order anisotropy field by $\mu_0 H_{a2} = \frac{2K_2}{M_S} = \frac{2K_{2\perp}}{M_S} - \mu_0(N_\perp - N_\parallel)M_S$ and fourth-order cubic anisotropy field by $\mu_0 H_{a4} = 2K_4/M_S$, here M_S is saturation magnetization, $K_{2\perp}$ is uniaxial out-of-plane anisotropy constant, K_4 is fourth order cubic anisotropy constant and $N_\perp$, $N_\parallel$ are the demagnetizing factors parallel and perpendicular to the film plane respectively. Transmission electron microscopy was performed using a JEOL 4000EX microscope with an acceleration voltage of 400 kV.

3. Results and discussion

In (Ge,Mn) films grown on Ge(001) substrate, Mn-rich nanocolumns spanning the whole film thickness are observed as a consequence of two-dimensional spinodal decomposition [4, 11]. These nanocolumns are well crystalline and in perfect epitaxial relationship with the Ge matrix when grown at substrate temperature of 100°C. However if the growth temperature is increased, they starts to become amorphous and increase in diameter. The high resolution TEM images of these nanocolumns are shown in reference [4, 11]. If we assume that the Mn concentration in the Ge matrix is in between 0 and 1 %, the resulting Mn content in the columns is close to 50 %.

SQUID measurements performed on a $Ge_{0.9}Mn_{0.1}$ film having crystalline and amorphous (Ge,Mn) nanocolumns are shown in Fig. 1. We identify two different contributions: paramagnetic signal from Mn ions in the Ge matrix at lower temperatures and ferromagnetic signal from the nanocolumns. The curie temperature was found to be 150K for crystalline nanocolumns and 125K for amorphous nanocolumns. The ZFC-FC measurements show that crystalline nanocolumns are superparamagnetic with blocking temperature of 15±5K and the narrow shape of the ZFC peak is related to the narrow size distribution of nanocolumns grown in this temperature range [11]. The amorphous nanocolumns have a Curie temperature of 125 K. In ZFC-FC curves we see two peaks at around 27 K and 72 K corresponding to broken and continuous nanocolumns. These peaks are broad and in agreement with the broad size distribution observed by TEM [11]. For these nanocolumns, the susceptibility perpendicular to

Journal of Physics: Conference Series **292** (2011) 012011 doi:10.1088/1742-6596/292/1/012011

the plane seems slightly higher than that in the plane but a strong diamagnetic signal from the substrate and paramagnetic signal from diluted Mn atoms makes it difficult to quantify precisely this anisotropy.

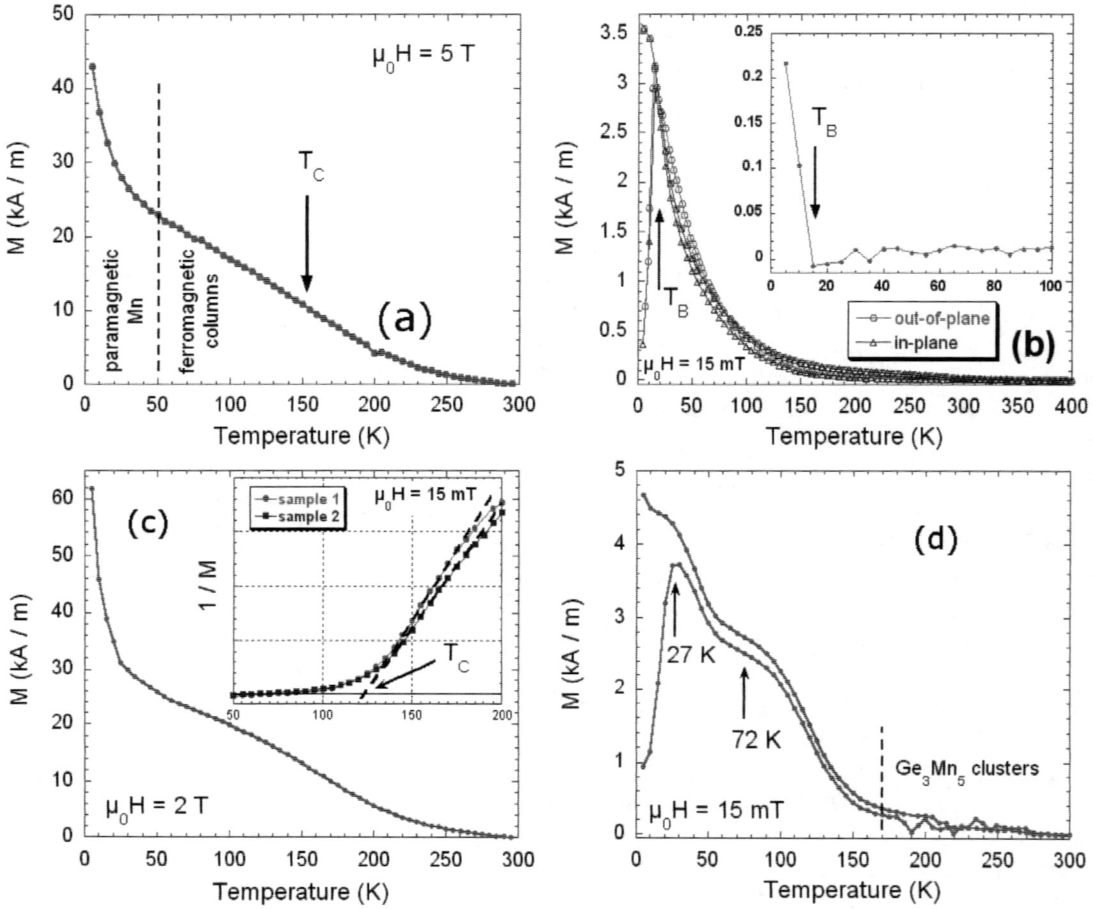

Figure 1. For T_g=100°C: (a) Temperature dependence of the saturation magnetization (in kA/m) of a $Ge_{0.9}Mn_{0.1}$ film containing crystalline nanocolumns. The applied field is 5 T. (b) ZFC-FC measurements performed with the field of 15 mT parallel and perpendicular to the film plane. Inset: magnetic remanence after maximum field cooling under 5 T. For T_g=150°C: (c) Temperature dependence of the saturation magnetization of a $Ge_{0.9}Mn_{0.1}$ film containing amorphous nanocolumns. The magnetic field is 5 T and applied in the film plane along the [110] direction. The Curie-Weiss plot for two identical samples is shown in the inset. (d) ZFC-FC measurements performed at 0.015 T. The ZFC-FC curve exhibits two peaks with blocking temperatures around 27 K and 72 K respectively.

EPR technique is hence used to differentiate the ferromagnetic signal of nanocolumns from the paramagnetic diluted Mn atoms and estimate magnetic anisotropy. The experiments were performed in Q-band (34 GHz) and the field was applied in out-of-plane geometry from the direction [110] to $[\bar{1}\bar{1}0]$ direction. Fig. 2(a,b) shows the EPR spectra of a $Ge_{0.9}Mn_{0.1}$ film with crystalline nanocolumns and the angular dependence of the resonance field is fitted using the Smit-Beljers formalism [19]. We found the anisotropy fields: $\mu_0 H_{a2}$ ≈0.09T for the second order and $\mu_0 H_{a4}$ ≈-0.11T for the fourth order. Hence, these nanocolumns exhibit a perpendicular uniaxial anisotropy and a cubic anisotropy with easy axis along [111]. If we consider the saturation magnetization of these small nanocolumns M_S=140±20 kA/m, we can derive the anisotropy constants to be: $K_2 \approx 0.63 \times 10^4 J/m^3$ and $K_4 \approx 0.77 \times 10^4 J/m^3$. The presence of cubic anisotropy supports the crystallinity of nanocolumns and the second order anisotropy well corresponds to shape magnetic anisotropy along the columns axis. The EPR spectra from amorphous columns and its angular dependence are shown in Fig. 2(c,d). A single perpendicular uniaxial anisotropy is found and the fit gives the corresponding anisotropy field to be: $\mu_0 H_{a2} \approx 0.09$ T. From the saturation magnetization of amorphous nanocolumns $M_S = 220 \pm 20$ kA/m, the anisotropy constant is calculated to be $K_2 \approx 1.1 \times 10^4 J/m^3$. This value corresponds to that of shape anisotropy $\mu_o M_S^2/4 = 1.5 \times 10^4 J/m^3$ of nanocolumns. As expected for amorphous nanocolumns, no cubic magneto-crystalline contribution is found.

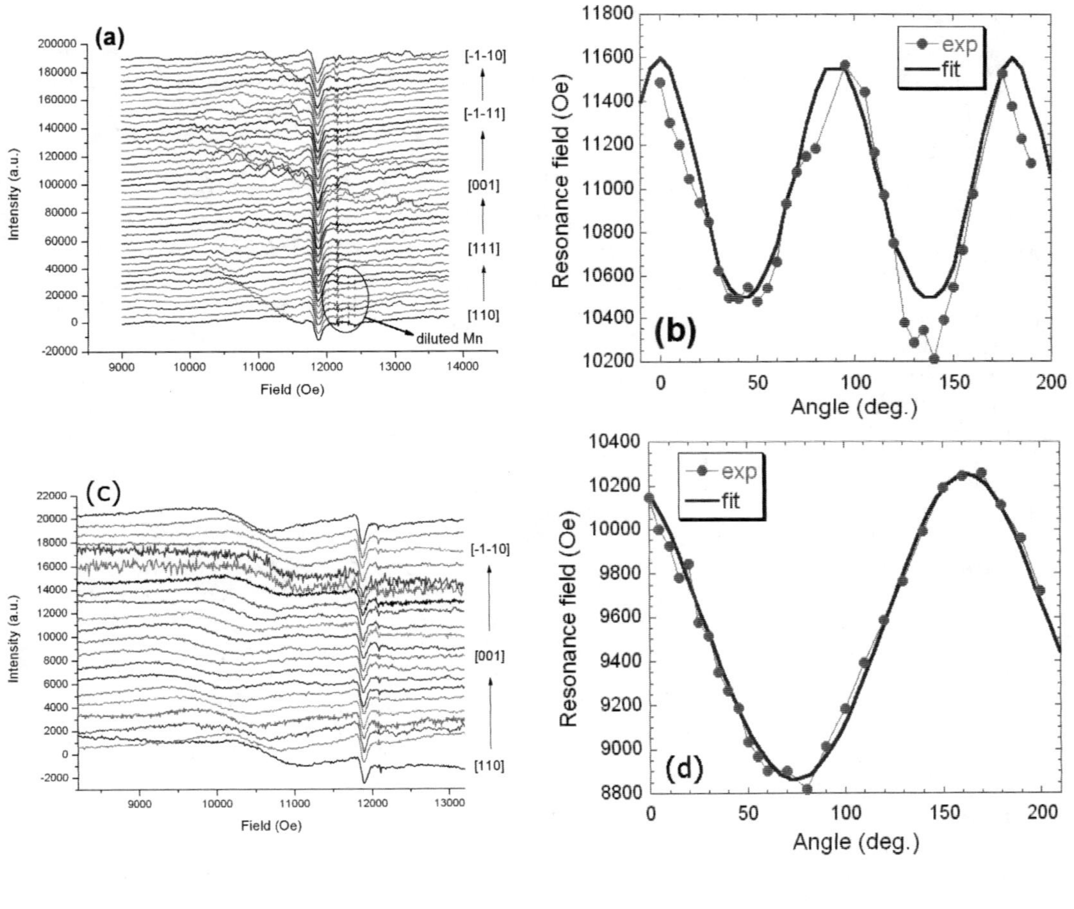

Figure 2. For T_g=100°C:(a) The EPR absorption spectra at Q band (34 GHz) of $Ge_{0.9}Mn_{0.1}$ sample. (b)Angular dependence of the resonance field as a function of orientation of magnetic field. We observe two contributions: a ferromagnetic peak of weak anisotropy and six hyperfine peaks from diluted paramagnetic Mn ions. For For T_g=150°C: (c) The EPR absorption spectra at Q band (34GHz) of $Ge_{0.9}Mn_{0.1}$ sample containing amorphous nanocolumns. (d)Angular dependence of the resonance field as a function of orientation of magnetic field. (The angle is defined between the direction [110] and the applied magnetic field. The angular step in goniometer during measurement has been 5°).

4. Conclusion

In this paper, we have systematically investigated different (Ge,Mn) nanostructures. We find a good correlation between their structural morphology and the magnetic properties.

In Ge_3Mn_5 films epitaxially grown on Ge(111), shape anisotropy is dominating and overcomes the magnetocrystalline component. Furthermore crystalline (Ge,Mn) nanocolumns exhibit second-order perpendicular uniaxial anisotropy along with fourth-order cubic anisotropy along (111) direction. This cubic anisotropy has been expected because of the crystalline structure of nanocolumns observed in TEM. On the other hand, in amorphous nanocolumns grown at high temperature (T_g =150°C), this cubic anisotropy is absent and only shape anisotropy is observed. The correlation between magnetization and the diamond lattice in these Mn-rich structures has never been evidenced before.

To conclude, the knowledge of magnetic anisotropy is important for future use of (Ge,Mn) material spintronics applications since it determines the direction of magnetization, coercive fields and domain sizes. For this purpose, using complementary experimental techniques has proven to be effective for investigating the magnetic properties of these nanostructures.

5. Acknowledgments

The authors would like to thank Dr. Joel Cibert for his significant inputs. The authors finally acknowledge the financial support from the French Agence Nationale pour la Recherche (ANR project GeMO) and the Nanoscience Foundation of Grenoble (RTRA project IMAGE).

6. References

[1] Žutić I, Fabian J and Das Sarma S 2004 *Rev. Mod. Phys.* **76** 323–410
[2] Dietl T 2002 *Semicond. Sci. Technol.* **17** 377
[3] Ohno H 1998 *Science* **281** 951
[4] Jamet M, Barski A, Devillers T, Poydenot V, Dujardin R, Bayle-Guillemaud P, Rothman J, Bellet-Amalric E, Marty A, Cibert J *et al.* 2006 *Nature Mater.* **5** 653–659
[5] Stroppa A, Picozzi S, Continenza A and Freeman A J 2003 *Phys. Rev. B* **68** 155203
[6] Park Y D, Hanbicki A T, Erwin S C, Hellberg C S, Sullivan J M, Mattson J E, Ambrose T F, Wilson A, Spanos G and Jonker B T 2002 *Science* **295** 651–654
[7] Bougeard D, Ahlers S, Trampert A, Sircar N and Abstreiter G 2006 *Phys. Rev. Lett.* **97** 237202
[8] Ottaviano L, Passacantando M, Verna A, Gunnella R, Principi E, Cicco A D, Impellizzeri G and Priolo F 2006 *Journal of Applied Physics* **100** 063528 (pages 4)
[9] Jaeger C, Bihler C, Vallaitis T, Goennenwein S T B, Opel M, Gross R and Brandt M S 2006 *Phys. Rev. B* **74** 045330
[10] Li A P, Zeng C, van Benthem K, Chisholm M F, Shen J, Nageswara Rao S V S, Dixit S K, Feldman L C, Petukhov A G, Foygel M and Weitering H H 2007 *Phys. Rev. B* **75** 201201
[11] Devillers T, Jamet M, Barski A, Poydenot V, Bayle-Guillemaud P, Bellet-Amalric E, Cherifi S and Cibert J 2007 *Phys. Rev. B* **76** 205306
[12] Pinto N, Morresi L, Ficcadenti M, Murri R, D'Orazio F, Lucari F, Boarino L and Amato G 2005 *Phys. Rev. B* **72** 165203
[13] Ahlers S, Bougeard D, Sircar N, Abstreiter G, Trampert A, Opel M and Gross R 2006 *Phys. Rev. B* **74** 214411
[14] Morgunov R, Farle M, Passacantando M, Ottaviano L and Kazakova O 2008 *Phys. Rev. B* **78** 045206
[15] Sugahara S, Lee K L, Yada S and Tanaka M 2005 *Jpn. J. Appl. Phys.* **44** L1426
[16] Sato K, Katayama-Yoshida H and Dederichs P H 2005 *Jpn. J. Appl. Phys.* **44** L948
[17] Zheng S, Zhu W, Gao Y F, Stocks G M and Zhang Z 2010 *Applied Physics Letters* **96** 071913 (pages 3)
[18] Hillebrands B and Ounadjela K (eds) 2002 *Spin Dynamics in Confined Magnetic Structures I* (Springer Berlin)
[19] Smit J and Beljers H 1955 *Phillips Res. Rep.* **10** 113
[20] Suhl H 1955 *Phys. Rev.* **97** 555
[21] Liu X and Furdyna J K 2006 *Journal of Physics: Condensed Matter* **18** R245
[22] Farle M 1998 *Reports on Progress in Physics* **61** 755
[23] Jain A, Jamet M, Barski A, Devillers T, Yu I S, Porret C, Bayle-Guillemaud P, Favre-Nicolin V, Gambarelli S, Maurel V *et al.* 2010 *Phys. Rev. B* **submitted**

The effects of Mn concentration on structural and magnetic properties of $Ge_{1-x}Mn_x$ diluted magnetic semiconductors

Thi-Giang Le[1, 3], D N H Nam[2], M-T Dau[1], T K P Luong[3], N V Khiem[3], V Le Thanh[1,*], L Michez[1] and J Derrien[1]

[1] Centre Interdisciplinaire de Nanoscience de Marseille (CINaM-CNRS), Aix-Marseille Université, Campus de Luminy, case 913, 13288 Marseille, France
[2] Department of Materials Science and Engineering, University of Virginia, 395 McCormick Road, Charlottesville, VA 22904, USA
[3] Department of Engineering and Technology, Hong Duc University, 307 Le Lai Street, Thanh Hoa City, Vietnam

*E-mail: lethanh@cinam.univ-mrs.fr

Abstract. Reflexion high-energy electron diffraction (RHEED), transmission electron microscopy (TEM) along with physical property measurement system (PPMS) were used to investigate the growth kinetics of $Ge_{1-x}Mn_x$ diluted magnetic semiconductors (DMS) grown on Ge(001) by means of molecular beam epitaxy (MBE). At a given intermediate growth temperature of 130 °C, we have identified the formation of successive heterogeneous phases when increasing the Mn concentration from 1 to 14 %: DMS phase containing nanosized Mn-rich clusters for x below 2%, DMS phase containing high Curie temperature (T_C) nanocolumns for x ranging from 5 to 6 %, DMS phase in which GeMn nanocolumns and Mn_5Ge_3 clusters coexist and then finally DMS containing mainly Mn_5Ge_3 clusters at Mn concentration higher than 12%. Our results confirm that the low solubility of Mn in Ge is the main origin of the formation of heterogeneous phases and provide evidence that it is extremely difficult to form a homogenous GeMn DMS even for Mn concentrations being below 2%. We also demonstrate that high-T_C nanocolumns and Mn_5Ge_3 clusters are competing processes and the process window corresponding to the stabilisation of high-T_C nanocolumns remains extremely tight.

1. Introduction

In recent years, diluted magnetic semiconductors (DMS), obtained by doping transition metals (TM) such as Mn, Ni, Fe, or Co into a host semiconducting matrix, have attracted great attention for their potential in combining ferromagnetic and semiconducting properties in a single material [1-3]. Among various materials, Ge:Mn-based DMS appears to be a promising candidate due to its compatibility with mainstream silicon technology, allowing a direct integration of future spintronic devices into the actual CMOS technology. For device applications, it is imperative to get DMS with Curie temperature being at least equal or above room temperature. This explains why up to now a great deal of interest has been focused on the investigation of the dependence of the Curie temperature of GeMn DMS on the growth parameters [4-10]. Among numerous growth parameters which can affect the GeMn growth behaviour, the growth temperature and the Mn concentration appear to be the most important parameters. Regarding the effect of * the growth temperature, in order to overcome the low thermodynamic solubility of Mn in Ge, it appears natural to carry out growth at temperatures low enough to bring the system far from equilibrium. According to previous works, three main regions of temperatures have been identified: for temperatures above 180 °C, intermetallic Mn_5Ge_3 precipitates

Journal of Physics: Conference Series **292** (2011) 012012 doi:10.1088/1742-6596/292/1/012012

or clusters are often observed [11-17] and in some cases their magnetic properties entirely dominate the magnetic signature of the whole film; For temperatures below 80 °C, Mn-rich elongated structures and even amorphous Mn-rich precipitates are shown to be formed [18-20]. Of particular interest, in the intermediate temperature range between 110 to 150 °C, it has been shown that GeMn nanocolumns with Curie temperature higher than 400K can be stabilized [21, 22]. However, little is known about the effect of the Mn concentration on sequential Ge:Mn phase formation [23].

In this work, we report on the kinetics of phase formation of $Ge_{1-x}Mn_x$ films on Ge(001) as a function of the Mn concentration, ranging from 1 to 14 %. We have chosen an intermediate substrate temperature of 130 °C to avoid a direct formation of intermetallic Mn_5Ge_3 precipitates produced at high temperatures [17] or amorphous Mn-rich clusters at temperatures below 80 °C [20]. We have identified sequential phase transitions when increasing Mn content and such a phase evolution can be attributed to two effects: a low Mn solubility and a Mn segregation toward the film surface. In particular, we show that the formation of high-T_C nanocolumn phase and intermetallic Mn_5Ge_3 clusters are two processes which are in competition and the process window for stabilisation of the nanocolumn phase is extremely tight.

2. Experimental

$Ge_{1-x}Mn_x$ films were grown by molecular beam epitaxy (MBE) on epi-ready *n*-type Ge(001) wafers with a nominal resistivity of 10 Ω.cm. The base pressure in the MBE system is better than 5×10^{-10} Torr. The growth chamber is equipped with a reflexion high-energy electron diffraction (RHEED) technique to control the cleanness of the substrate surface prior to growth and to monitor the epitaxial growth process. $Ge_{1-x}Mn_x$ films were obtained by co-deposition of Ge and Mn from standard Knudsen effusion cells, the Ge deposition rate was determined from RHEED intensity oscillations whereas the Mn deposition rate was deduced from Rutherford backscattering spectrometry (RBS) measurements. For Mn concentrations below 2%, the measurement uncertainty can reach a value of 10%. The standard growth rate used in this work is of 1 – 2 nm/min.

Structural analyses of the grown films were performed through extensive high resolution transmission electron microscopy (TEM) by using a JEOL 3010 microscope operating at 300 kV with a spatial resolution of 1.7 Å. The magnetic properties were investigated by using Quantum Design physical property measurement system (PPMS) in the temperature range from 5 to 350 K and with a magnetic field of 0.1T applied in the plane of samples. The diamagnetic contributions of Ge substrates were subtracted from the measurements, leaving the magnetic contributions of $Ge_{1-x}Mn_x$ films.
The cleaning of Ge surfaces was carried out in two steps: a chemical cleaning to remove hydrocarbon related contaminants followed by an *in-situ* thermal cleaning at ~750°C to remove the Ge surface oxide layers. After this step, the Ge(001) surface generally exhibits a (2x1) reconstruction. To insure a good starting Ge surface prior to $Ge_{1-x}Mn_x$ growth, a ~30 nm thick Ge buffer layer was systematically grown at a substrate temperature of 600°C.

3. Results and discussion

Figures 1(a) and (b) respectively display RHEED patterns taken along the [1-10] azimuth of the clean Ge surface prior to growth and during $Ge_{1-x}Mn_x$ growth with Mn concentration being in the range from 1 to 2%. Prior to growth, the Ge surface is characterized by a well-developed 2x1 streaky pattern, indicating that the surface is clean and smooth. This is also confirmed by the observation of high-intensity Kikuchi lines, which overlap 1x1 and specular streaks, giving rise to localized reinforcements of intensity, as can be seen in the RHEED pattern. Compared to Ge surface, $Ge_{1-x}Mn_x$ surface exhibits a faint pattern with ½ streaks of weak intensity. Furthermore, three-dimensional spots due to transmission diffraction become visible. This indicates that even for Mn concentration as low as 1-2%, $Ge_{1-x}Mn_x$ films are somewhat disordered and the surface becomes rough. To understand such behaviour, we present in figure 1(c) a typical cross-sectional TEM image of the corresponding sample.

Dark contrast observed between Ge substrate and buffer layers is due to carbon contamination of the initial substrate surface occurred when the cleaning was not successful, this confirms the necessity of growing Ge buffer layers prior to $Ge_{1-x}Mn_x$ deposition. The white dotted line was added to indicate the interface between Ge buffer layers and $Ge_{1-x}Mn_x$ film. The image clearly reveals that the $Ge_{1-x}Mn_x$ film is not homogenous at all but contains Mn-rich clusters. The clusters have an average size of 1-2 nm and are randomly distributed in the film. It is worth noting that even for a Mn concentration reduced down to 1%, it was not possible to get homogenous $Ge_{1-x}Mn_x$ films.

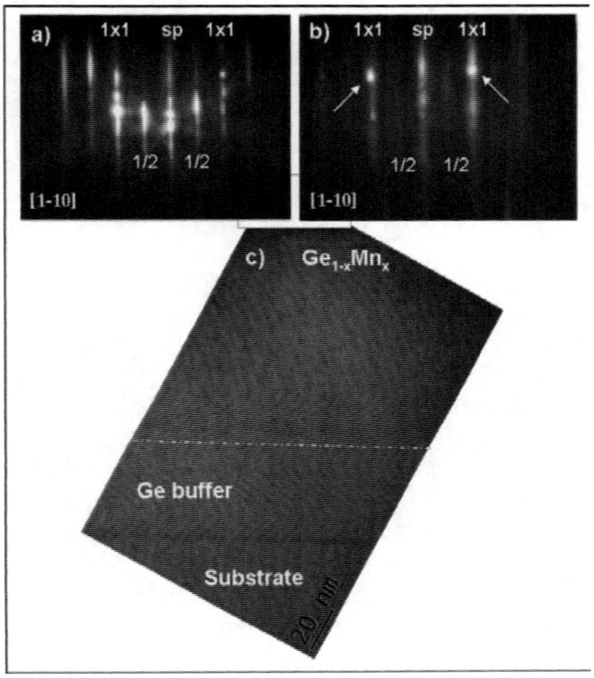

Figure 1. (a, b): RHEED patterns taken along [1-10] azimuth of the Ge(001) surface prior to $Ge_{1-x}Mn_x$ growth (a) and during $Ge_{1-x}Mn_x$ growth with x being in the range of 1-2% (b). The specular streak (sp) together with 1x1 bulk-like and half-ordered (1/2) streaks are indicated. Three-dimensional spots arising from a rough surface are indicated by small arrows in (b). Typical cross-sectional TEM image observed for a $Ge_{1-x}Mn_x$ film with x ~ 2% (c).

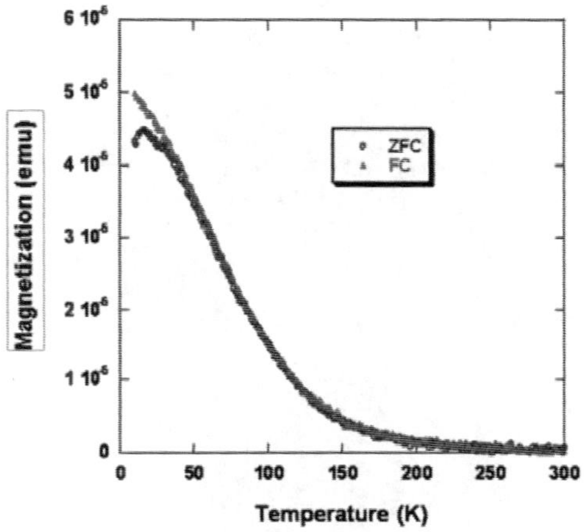

Figure 2. Typical evolution with temperature of magnetization of $Ge_{1-x}Mn_x$ with Mn concentration of ~2%. A Curie temperature of around 150K is deduced and can be attributed to the DMS matrix. The superimposition of ZFC and FC curves suggests that nanosized Mn-rich clusters do not contribute to the magnetic properties of the grown film.

We show in figure 2 the temperature dependence of net magnetization (Ms) of a $Ge_{1-x}Mn_x$ film with x ~2%. The magnetization exhibits a monotonous and smooth decrease with increasing the temperature, suggesting that it arises from an unique phase. By making derivation of Ms versus temperature, a T_C of ~150 K is deduced, a value expected for a $Ge_{1-x}Mn_x$ alloy with 2% of Mn. This suggests that the nanosized Mn-rich clusters dot not contribute to the film magnetization, which is mainly dominated by the diluted matrix. We also note that for a Mn concentration of ~1%, a Curie temperature of only 80 K was observed.

When the Mn concentration increases to 5 - 6%, the Mn segregation has organized in a specific manner that conducts to the formation of nanocolumns. Figure 3 shows the surface morphology observed by RHEED and an overview of the nanocolumn formation by cross-sectional TEM. The sample surface has become rougher, probably due to the presence of nanocolumns. TEM image shown in figure 3(b) reveals that most of nanocolumns are oriented perpendicular to the film, along the growth direction. The average size of nanocolumn is about 3-5 nm and no presence of Mn_5Ge_3 clusters are detectable. These results are very similar to those previously reported, in which $Ge_{1-x}Mn_x$ nanocolumns exhibit a Curie temperature well above 400 K [21, 22]. Our magnetic characterizations also confirm a Curie temperature higher than 350 K, the upper limit of temperature measurements in our PPMS. The authors of refs [21, 22] have attributed the composition of nanocolumns close to Ge_2Mn, an unknown Ge-rich phase in the Ge-Mn phase diagram. Recently, we use Laser Assisted Atom Probe Tomography (LA-APT) technique to investigate the Mn concentration inside nanocolumns and also in the matrix in-between nanocolumns. The obtained results, which will be published in a separate paper, clearly demonstrate that the Mn concentration in nanocolumns is neither constant nor close to Ge_2Mn.

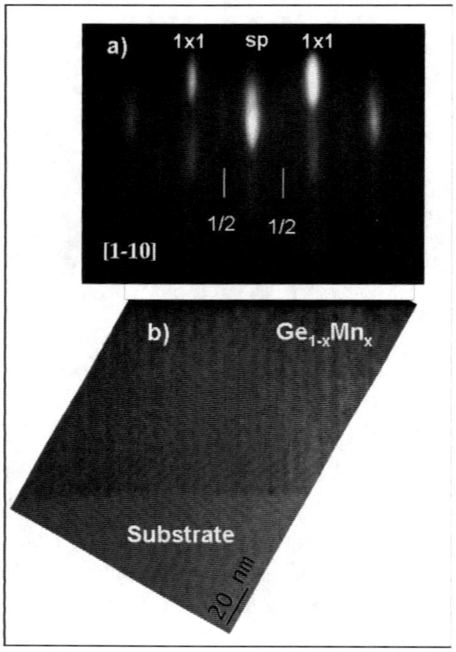

Figure 3. (a): RHEED pattern taken along [1-10] azimuth during $Ge_{1-x}Mn_x$ growth with x in the range of 5-6%. The specular streak (sp) together with 1x1 bulk-like and half-ordered (1/2) streaks are indicated.
(b): Typical cross-sectional TEM image observed for $Ge_{1-x}Mn_x$ film with x in the range of 5 – 6 %. The image reveals the formation of GeMn nanocolumns, which are separated by a DMS matrix less rich in Mn. Note that most of nanocolumns are oriented along the growth direction and perpendicular to the film layer.

With further increase of Mn up to ~14%, our results reveal a particular mechanism along which the formation of nanocolumns and Mn_5Ge_3 clusters is in competition. These results are depicted in figure 4: figures 4(a) and 4(b) report on the main features observed when the Mn concentration is in the range of 8 – 10 % whereas figure 4(c) and 4(d) correspond to Mn of 12 – 14 %. First of all, looking at the evolution of RHEED patterns, it can be clearly seen that the film surface roughness greatly increases at this Mn concentration range. For Mn concentration of 8 – 10 %, streaky features arising from reflexion diffraction have completely disappeared, only three-dimensional transmission diffraction is observed. When the Mn concentration reaches 12 – 14 %, diffraction patterns become so faint that polycrystalline features starts to appear.

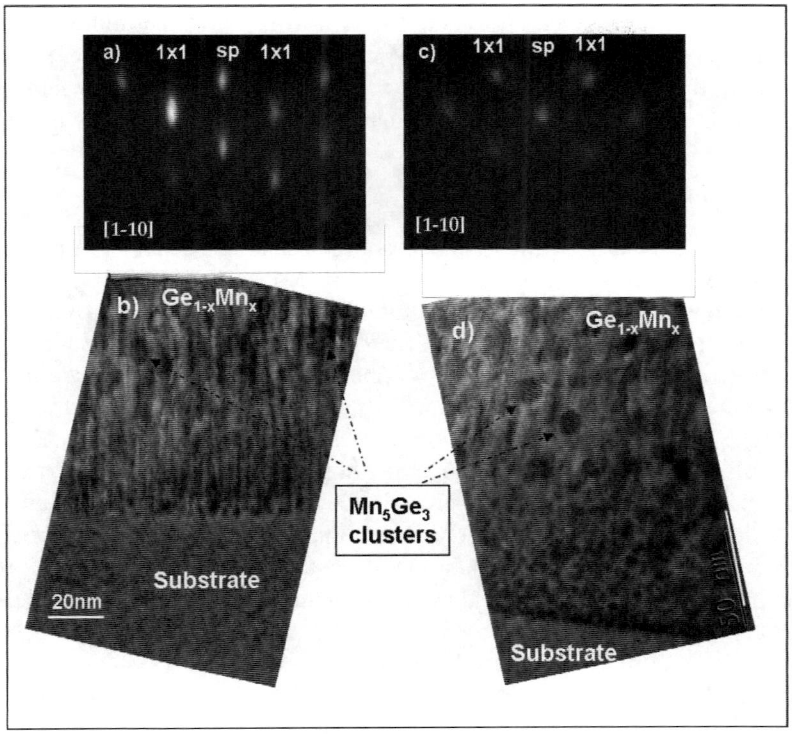

Figure 4. RHEED patterns (a, c) and cross-sectional TEM images (b and d) of $Ge_{1-x}Mn_x$ film with high Mn concentrations.
- (a, b) for Mn concentration ranging from 8 to 10 %.
- (c, d) for Mn concentration ranging from 12 to 14 %.

Figures 4(b) and 4(d) provide an interesting view of the competition between $Ge_{1-x}Mn_x$ nanocolumns and Mn_5Ge_3 clusters. In figure 4(b), we still observe the presence of nanocolumns in the region near the interface (for thickness up to 30 nm starting from the interface). Going up to the film surface, one can recognize the presence of some Mn_5Ge_3 clusters and at the same time nanocolumns become disrupted and somewhat disordered. To understand such an evolution, it is important to recall that in $Ge_{1-x}Mn_x$ films, Mn_5Ge_3 clusters are a stable phase, they are formed each time the Mn concentration and/or the growth temperature become high enough whereas $Ge_{1-x}Mn_x$ nanocolumns are metastable, they transform into Mn_5Ge_3 clusters for thermal annealing at around 400 °C. The above phase transition can be explained as followed: at this growth condition of the substrate temperature of 130 °C and Mn concentration of 8 – 10 %, the formation of $Ge_{1-x}Mn_x$ nanocolumns is still favorable. However, the Mn concentration inside nanocolumns should be high and in addition it increases along nanocolumns toward the film surface due to Mn segregation when the film thickness increases. When the Mn concentration reaches a maximal value of about 62 – 63 %, which is equal to Mn concentration in Mn_5Ge_3 compound, nanocolumns become metastable, resulting to the formation of Mn_5Ge_3 clusters in the region close to the film surface. Such an explanation also provides a good interpretation of the phase formation when the Mn concentration increases to 12 - 14 %. At this Mn concentration range, a direct formation of Mn_5Ge_3 clusters becomes thermodynamically more favorable than nanocolumns. As can be seen in figure 4(d), no presence of $Ge_{1-x}Mn_x$ nanocolumns is visible even in the interface region. Mn_5Ge_3 clusters of very small size are formed at the interface just after the deposition of the first layers. The size of Mn_5Ge_3 clusters then increase with increasing the film thickness, due to Mn segregation toward the film surface. We also note that magnetization measurements (not shown here) confirm the coexistence of two Curie temperatures: one at ~304 K, which can be attributed to Mn_5Ge_3

clusters [24] and the other observed at around 107 K, arising from the diluted matrix. This value of 107 K is smaller that T_C observed for a $Ge_{1-x}Mn_x$ alloy with 2% of Mn because the formation of Mn_5Ge_3 clusters are thermodynamically more favorable and it occurs by consuming Mn beforehand incorporated in substitutional sites of the diluted matrix.

4. Conclusion

In conclusion, we have chosen an intermediate and appropriate substrate temperature of 130 °C to investigate the sequential phase formation in $Ge_{1-x}Mn_x$ DMS as a function of the Mn concentration. We show that the low solubility of Mn is the main obstacle to form homogenous diluted magnetic semiconductors. Even for Mn concentration as low as 1%, we show that it is unable to form homogenous diluted materials. It should be possible to get homogenous materials for a lower Mn concentration but the Curie temperature will be very low. We have identified transition of successive $Ge_{1-x}Mn_x$ DMS phases when increasing the Mn concentration: DMS phase containing nanosized Mn-rich clusters for x below 2%, DMS phase containing high-T_C nanocolumns for x ranging from 5 to 6 %, DMS phase in which GeMn nanocolumns and Mn_5Ge_3 clusters coexist and finally DMS containing mainly Mn_5Ge_3 clusters for Mn concentrations higher than 12 %. Obviously, the separation boundaries of these phases are not sharp, they depend on the Mn concentration and interestingly, also on the film thickness. We have provided evidence that the formation of high-T_C nanocolumns and Mn_5Ge_3 clusters are competing process and the process window for stabilizing the high-T_C nanocolumn phase is very small. It is important to precisely control not only the Mn concentration but also the corresponding film thickness before nanocolumns can be transformed to Mn_5Ge_3 clusters due to Mn segregation toward the film surface.

References

[1] Ohno H 1998 *Science* **281** 951
[2] Dietl T *Semicond. Sci. Technol.* **17** 377
[3] Stroppa A, Picozzi S, Continenza A and Freeman A J 2003 *Phys. Rev. B* **68** 155203
[4] Pinto N, Morresi L, Ficcadenti M, Murri R, D'Orazio F, Lucari F, Boarino L and Amato G 2005 *Phys. Rev. B* **72** 165203
[5] Tsui F, He L, Ma L, Tkachuk A, Chu Y S, Nakajima K and Chikyow T 2003 *Phys. Rev. Lett.* **91** 177203
[6] Braak H et al. 2005 *J. Magn. Magn. Mater.* **286** 46
[7] Morresi L, Ayoub J, Pinto N, Ficcadenti M, Murri R, Ronda A and Berbezier I 2006 *Mater. Sci. Semicond. Process.* **9** 836
[8] Li A P, Wendelken J F, Shen J, Feldman L C, Thompson J R, Weitering H H 2005 *Phys. Rev. B* **72** 195205
[9] Cho S, Choi S, Hong S C, Kim Y, Ketterson J B, Kim B J, Kim YC and J.-H. Jung, *Phys. Rev. B* **66** (2002) 033303
[10] Park Y D, Hanbicki A T, Erwin S C, Hellberg C S, Sullivan J M, Mattson J E, Ambrose T F, Wilson A, Spanos G and Jonker B T 2002 *Science* **295** 651
[11] Park Y D, Wilson A, Hanbicki A T, Matteson J E, Ambrose T, Spanos G and Jonker B T 2001 *Appl. Phys. Lett.* **78** 2739
[12] Li A P, Shun J, Thompson J R and Weitering H H 2005 *Appl. Phys. Lett.* **86** 152507
[13] Ottaviano L, Passacantando M, Picozzi S, Continenza A, Gunnella R, Verna A, Impellizzeri G and Priolo F 2006 *Appl. Phys. Lett.* **88** 061907
[14] Passacantando M, Ottaviano L, D'Orazio F, Lucari F, De Biase M, Impellizzeri G and Priolo F 2006 *Phys. Rev. B* **73** 195207
[15] Verna A et al. 2006 *Phys. Rev. B* **74** 085204
[16] Chen J, Wang K L and Galatsis K 2007 *Appl. Phys. Lett.* **90** 012510

[17] Bihler C, Jaeger C, Vallaitis T, Gjukic M, Brandt M S, Pippel E, Woltersdorf J and Gosele U 2006 *Appl. Phys. Lett.* **88** 112506
[18] Ahlers S, Bougeard D, Sircar N, Abstreiter G, Trampert A, Opel M and R Gross 2006 *Phys. Rev. B* **74** 214411
[19] Bougeard D, Ahlers S, Trampert A, Sircar N and Abstreiter G 2006 Phys. Rev. Lett. **97** 237202
[20] Bougeard D, Sircar N, Ahlers S, Lang V, Abstreiter G, Trampert A, LeBeau J M, Stemmer S, Saxey D W and Cerezo A 2009 *NanoLetters* **9** 3743
[21] Jamet M et al. 2006 *Nat. Mater.* **5** 653.
[22] Devillers T, Jamet M, Barski A, Poydenot V, Bayle-Guillemaud P, Bellet-Amalric E, Cherifi S and Cibert J 2007 *Phys. Rev. B* **76** 205306.
[23] Li H, Wu Y, Guo Z, Luo P and Wang S 2006 *J. Appl. Phys.* **100** 103908
[24] Tawara Y and Sato V 1963 *J. Phys. Soc. Jpn.* **18** 773

Understanding the Room Temperature Ferromagnetism in GaN Nanowires with Pd Doping

Sujit Manna* **and S. K. De**

Department of Materials Science, Indian Association for the Cultivation of Science, Jadavpur; Kolkata 700 032; INDIA

E-mail: mannaju@gmail

Abstract. We report the first synthesis and characterization of 4d transition metal palladium-doped GaN nanowires (NWs). Room temperature ferromagnetism has been observed in high quality Vapor Liquid Solid (VLS) epitaxy grown undoped n-type GaN nanowires. It was proposed that this type of magnetism is due to defects which are not observed in Bulk GaN because of large formation energy of defects in bulk GaN. Here we have successfully doped 4d transition metal Pd in GaN NWs. We find fairly strong and long-range ferromagnetic coupling between Pd substituted for Ga in GaN . The results suggest that 4d metals such as Pd may also be considered as candidates for ferromagnetic dopants in semiconductors.

1. Introduction

Doping semiconductor nanostructures with selected dopant ions represents an effective means of imparting new electrical[1], optical[2, 3], and magnetic[4], properties into these promising materials, thereby expanding their intrinsic functionalities. In particular, nanostructured diluted magnetic semiconductors (DMSs)[5], obtained by substitutional doping of nanocrystalline compound semiconductors with paramagnetic transition-metal ions, can enable simultaneous control and manipulation of electron spins and charges at the nanoscale. This approach is pivotal in the formulation of spin-based electronics, or spintronics, which has emerged as a potential alternative to conventional charge-based electronics[6]. Diluted magnetic semiconductor nanowires (DMS-NWs)[2, 7, 8] have, therefore, attracted significant interest in recent years due to the possibility of expanding devices functionality using electronic spin as an additional degree of freedom. Moreover, shape anisotropy of DMS-NWs may potentially allow for a reduced magnetostatic energy and, therefore, easier magnetization of the NWs along the growth direction. Considerable effort has gone so far into the study of 3d transition metals, such as manganese and chromium as dopants in GaN nanowires . As far as 4d metals are concerned, only the optical spectra of niobium and molybdenum impurities in GaN have been considered thus far. It has previously been shown that palladium, which is nonmagnetic in its fcc ground state, does exhibit ferromagnetic order when in a hcp or double-hcp (dhcp) phase[9]. A Pd atom in these phases experiences an environment of similar symmetry as in a wurtzite semiconductor such as gallium nitride.

Herein, we synthesized high-purity single-crystalline $Ga_{1-x}Pd_xN$ NWs with controlled Pd contents, $x = 0$, 0.0049, 0.0076 and 0.016 exhibiting room-temperature ferromagnetism, using a simple vapor transport method. We attempted to correlate the magnetization of these nanowires with their electronic structures.

2. Experimental

Pure and Palladium doped GaN NWs were prepared via metal-catalyzed Vapor Liquid Solid (VLS) method using ammonia (99.99%, Matheson), gallium metal (99.99%, Aldrich), $PdCl_2$ (99.99%,Aldrich) as the N, Ga and Pd sources respectively, and phosphorus doped n-type Si (100) as the growth substrate. First bulk GaN precursor was prepared by dissolving gallium metal in 69% HNO_3 (Aldrich) solution such that the Ga: HNO_3 molar ratio was greater than 1:3 under ultrasonication. This solution was then diluted by adding into 2-propanol under stirring such that the molar ratio of Ga:2-propanol was 1:90 and allowed to stir for 30 min. A white precipitate was obtained on slow addition of 25% NH_4OH (ammonia solution) to this solution under vigorous stirring, which was separated by centrifugation, washed in ethanol, and dried at 60°C for 24 h. Nitridation was carried out in an alumina boat kept in a quartz tube with an inner diameter equal to 24 mm under ammonia flow. The tube was flushed with nitrogen for 30-45 min, and ammonia was allowed to flow into the tube. The flow was set to around 200 sccm, and the temperature was fixed to 900°C for 60 min. A yellow color pure GaN bulk powder was obtained. These GaN powders are taken as a source material. GaN powders and $PdCl_2$ powders were placed separately in two alumina boats, which loaded inside a quartz tube reactor. A silicon substrate, on which a 3-5 nm thick Au film was deposited, was positioned at a distance of 10 cm away from the GaN source. Ammonia was allowed to flow at a rate 300 sccm while raising and lowering the temperature. The temperature of Ga and Pd sources were set to 1100°C and 800°C, respectively, and the substrate was approximately set at 900°C. Argon gas was continuously flowed at a rate of 500 sccm for 60 min during the synthesis. The Pd content was controlled by adjusting the evaporation temperature of $PdCl_2$. Undoped GaN nanowires was prepared under identical conditions, but without $PdCl_2$.

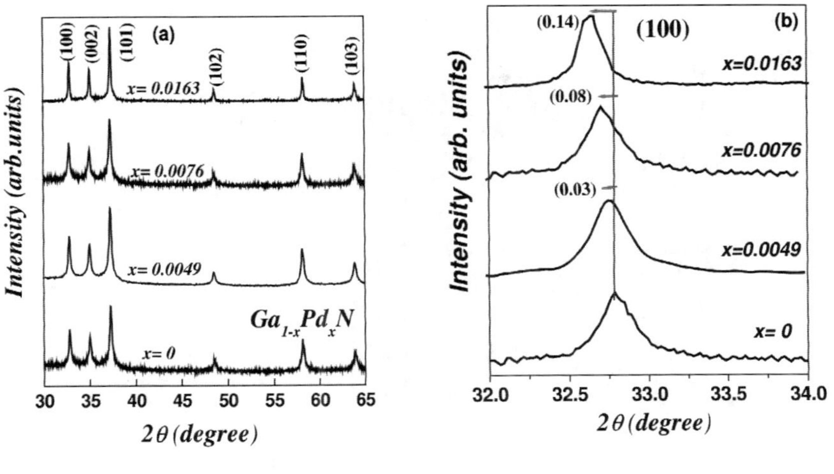

Figure 1. (a) Full range XRD patterns taken from $Ga_{1-x}Pd_xN$ (x= 0.0049, 0.0076, and 0.016) NWs and (b) the magnified scaled (100) peak.

X-ray diffraction patterns of the samples were recorded by high resolution X'Pert PRO Panalytical X-ray diffractometer in the range 30-65^0 using Cu K_α radiation. Microstructure and crystal structures of the nanowires were obtained using Field Emission Scanning Electron Microscope(FESEM, JEOL)Transmission Electron Microscope (TEM) and high-resolution TEM

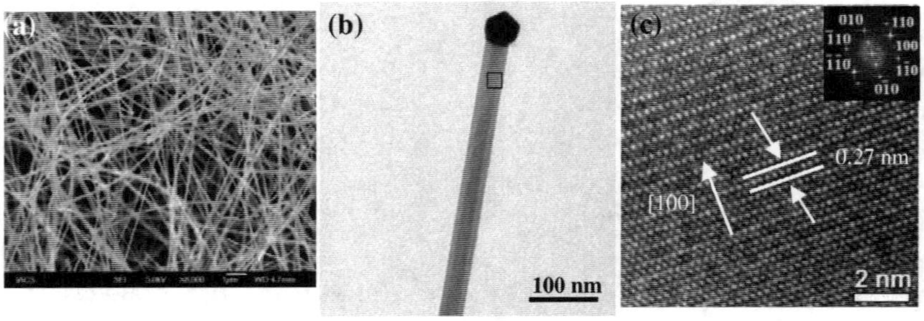

Figure 2. (a) Low magnified SEM micrograph of high density $Ga_{0.983}Pd_{0.016}N$ NWs, homogeneously grown on the substrate.(b) TEM image of single 40 nm diameter GaN nanowire that terminates in a faceted nanoparticle of higher contrast (c) HRTEM image of same nanowire. Inset shows a FFT pattern of HRTEM image

(HRTEM, JEOL 2010) studies. The metal contents in the nanowires were determined by an inductively coupled plasma atomic emission spectrometer (ICP-AES), (Perkin Elmer). Doped samples were characterized by X-ray photoelectron spectroscopy (XPS,5400 ESCA). Magnetization measurements were carried out by SQUID Magnetometer, Quantum Design, MPMS XL (evercool).

3. Results and Discussion

The high-resolution XRD patterns of the undoped GaN NWs and $Ga_{1-x}Pd_xN$ NWs are displayed in Figure 1a. The peaks of the GaN NWs exactly match those of wurtzite GaN. All of the GaN NW samples have a highly crystalline nature without the presence of other phases. Panels b in Figure 1 display the magnified (100) peaks.

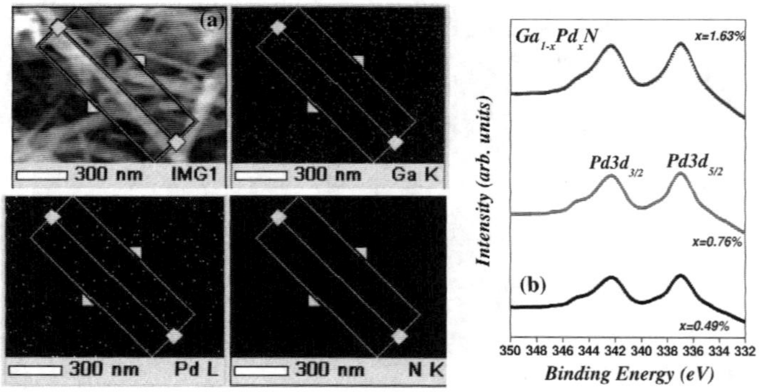

Figure 3. (a) EDS elemental mapping reveals the homogeneous composition of Ga, Pd and N. (b) Pd Core level XPS spectra for all Pd doped GaN NWs.

The peak position shifts to a lower angle most significantly with the Pd content. The maximum shifts 2Θ of the (100) peaks is 0.14. The lattice constant also increases with doping concentration of pd. As the Ga^{3+} ions ($r_{Ga} \sim 0.61$) at the tetrahedral sites were substituted with the larger radius Pd^{2+} ions ($r_{Pd} \sim 0.98$), the lattice constant would expand because of the difference in the ionic radius.This expansion of lattice constant can be ascribed to the hybridization between the Pd dopants and host defects (holes)

The growth mechanism of the nanowires follows the vapour-liquid-solid mechanism, which makes use of Au catalytic nanoparticles. Thermal decomposition of $PdCl_2$ occurs above 680^0C: $PdCl_2 = Pd + Cl_2$. Liquid Au droplet on Si substrate forms alloy with GaN and Pd. The supersaturation state of liquid alloy by absorbing vapor of constituent elements leads to nucleation at liquid-solid interface. The dopant Pd content was controlled by adjusting the evaporation temperature of Pd precursor. Figure 2a shows the low-magnified SEM micrograph of the high-density and uniform 1.63% Pd doped GaN($Ga_{0.983}Pd_{0.016}N$) NWs grown on the substrates. The TEM image of the individual GaN nanowire explicitly reveals their smooth surface and average diameter of 40 nm and 30-40 μm in length (Figure 2b). The lattice resolved TEM image and corresponding Fourier-transform electron diffraction (FFT ED) pattern of a selected nanowire reveal that it is composed of single-crystalline wurtzite structured GaN (Figure 2c and inset). It shows a spacing between neighboring (100) planes of ca. 0.27 nm (JCPDS Card no. 50-0792). The FFT ED pattern generated from the inversion of the TEM image using Digital Micrograph GMS1.2 software (Gatan Inc.) at the [0001] zone axis, confirms that it has the [1010] growth direction.

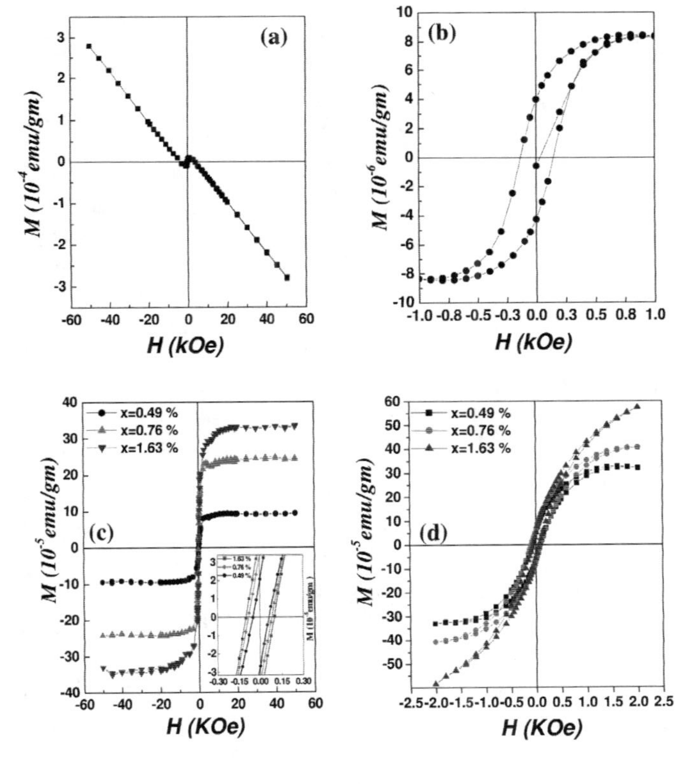

Figure 4. Magnetization (M) versus magnetic field (H) of pure GaN NWS at room temperature (a) without background correction. (b) After diamagnetic background correction. M vs H at 300 K (c) and 5 K (d) for all Pd doped GaN NWs. Inset shows the magnified M-H curves in the vicinity of H=0.

The EDS mapping for a selected $Ga_{0.983}Pd_{0.016}N$ nanowires is shown in Figure 3a. EDS line-scanning analysis indicates the Pd content (x= [Pd]/([Pd]+[Ga])) of the individual nanowire to be about x=0.02±0.01 with a homogeneous distribution being observed along the nanowires. The Pd content in the doped GaN NWs also determined from Inductive Couple Plasma Atomic Emission Spectroscopy(ICP-AES) with a high accuracy. The concentration of Palladium for highest doped sample as determined from ICP-AES is $\sim$ 1.63%. Electronic structure of Ga and dopant Pd is determined from XPS study. The fine scanned Pd3d core level spectra for all Pd doped GaN NWs are shown in Figure 3b. Which confirmed that in the sample palladium exists in Pd^{+2}. The peak of the nanowires appeares in the lower energy region compared to that of O(in PdO), due to the lower electronegativity of N. Therefore, the peaks would be expected to originate mainly from Pd(II)-N bonding structures. The carrier concentration and conductive type of these Pd-doped GaN nanowires were determined by Hall measurements at room temperature by Van der Pauw methods. The carrier concentration of Pd doped p-type GaN nanowires is 3.0×10^{18} cm^{-3} and a mobility of 118 cm^2 V^{-1} s^{-1}. It was suggested that, in p-type GaN, there exists a shallow acceptor's hole transfer process onto Pd^{+2} to form Pd^{+3}, where the Pd^{+3} ions act as effective shallow donors. This hole transfer process can be described as $Pd^{+2} + h^{+}$(acceptor)$\longrightarrow Pd^{+3}$. So there is a strong hybridization between the Pd dopant and the acceptor in the Pd doped GaN nanowires.

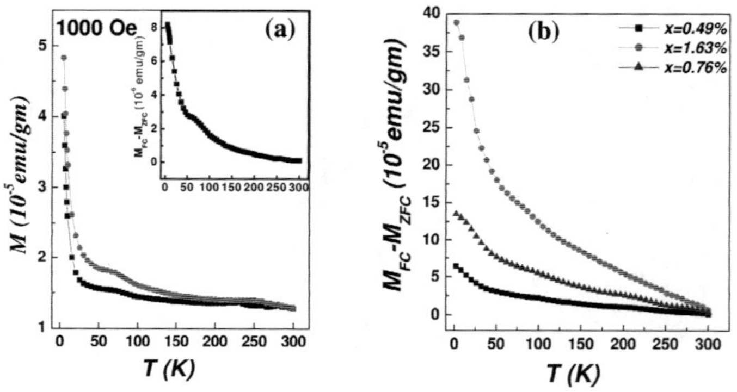

Figure 5. (a)Temperature dependent M_{FC} and M_{ZFC} with H=1kOe for undoped GaN NWs. The inset represents its ΔM (M_{FC} - M_{ZFC}) curve. (b) Temperature dependent ΔM (M_{FC} - M_{ZFC})curve for all doped sample.

The magnetic moment (M) versus magnetic field (H) curves for pure GaN NWs at 300 K are shown in Figure 4a. The samples exhibit ferromagnetic hysteresis behavior with a strong diamagnetic background signal. Figure 4b corresponds to the field-dependent M-H curve of the undoped GaN NWs after strong diamagnetic correction. Hysteresis occurs with a saturation field of 1 kOe. The saturated magnetization is estimated to be 8.5x10^{-6} emu/gm and coercivity 148 Oe. Figure 5a displays the field-cooled (M_{FC}) and zero-field cooled M (M_{ZFC}) versus T curves at H =1000 Oe for pure GaN NWs. The inset corresponds to ΔM (M_{FC} - M_{ZFC}) as a function of temperature in the range of 5-300 K. Figure 5b shows the same curve for all Pd doped sample. This magnetization difference is particularly informative when there are small amounts of ferromagnetic material in the presence of a large diamagnetic background. This subtraction indicates the presence of ferromagnetism if the difference is nonzero. This data indicates that the ferromagnetism persists to 320 K.

Both Ga and N vacancies are suggested to be present in the as-grown GaN depending on the growth conditions. Calculations of the energies of formation of Ga and N vacancies in bulk GaN have been reported in the literature[10]. Although the formation of nitrogen vacancy has lower energy, electronic structure calculations have shown that this type of vacancy causes a paramagnetic state. On the other hand, Ga vacancies introduce magnetic moments which lie on the neighboring N atoms that are spin polarized due to Hunds rule. This type of magnetism due to defects in bulk GaN has not been observed because of the large formation energy of defects in bulk GaN. However, the defect formation energy at the surface of the nanowires may be significantly different from that in the bulk due to the size effect which involves structural as well as electronic effects. Thus, the defect formation energies at the surface are lower, resulting in high concentration of defects which gives rise to percolative ferromagnetism at the surface of the nanowires of GaN.

Figure 4c-d displays the M-H curves of the $Ga_{1-x}Pd_xN$ NWs consistently showing hysteresis at 5k and 300 K. The saturated magnetizations of these NWs are estimated to be about 9.33x10^{-5} , 24.12x10^{-5} and 32.78x10^{-5}emu/gm at 300 K. The inset displays the curve in the vicinity of H=0. The hysteresis curves show that M_r =2.07x10^{-5}, 3.45x10^{-5} and 4.35x10^{-5} emu/gm and H_C= 60, 82, and 100 Oe for x= 0.0049, 0.0076, and 0.016 respectively. So as the Pd content increases the magnetization also increases. Theoretically it was predicted that the introduction of palladium introduces an impurity gap state, with both the majority and minority spin components retaining a band gap. Importantly, the Pd impurity related band is spin polarized, with a spin polarization of 0.24-0.31 eV and the Fermi level lying within this band; charge carriers traveling through a $Ga_{1-x}Pd_xN$ layer will therefore be spin polarized. This additional spin polarized state contribute to ferromagnetism significantly. As a result of it the magnetic moment of Pd doped GaN NWs is enhanced.

4. Conclusion

Single-crystalline $Ga_{1-x}Pd_xN$ NWs were grown by the vapor transport method, using the evaporation of GaN/$PdCl_2$. We have demonstrated that the undoped GaN NWs show room-temperature ferromagnetism that has been discussed in terms of defects at the surface of the nanowires. We find fairly strong and long range ferromagnetic coupling between Pd substituted for Ga in GaN. The results suggest that 4d metals such as Pd may also be considered as candidates for ferromagntic dopants in semiconductors.

5. Acknowledgments

The author Sujit Manna acknowledge the financial support of the Council of Scientific and Industrial Research (CSIR) and international travel support of Department of Science and Technology (DST), New Delhi, India.

6. References

[1] Zhong Z, Qian F, Wang D and Lieber C M 2003 *Nano Lett.* **3** 343
[2] Radovanovic P V, Barrelet C J, Gradecak S, Qian F and Lieber C M 2005 *Nano Lett.* **5** 1407
[3] Boyer J C, Vetrone F, Cuccia L A and Capobianco J A 2006 *J. Am. Chem. Soc.* **128** 7444
[4] Radovanovic P V and Gamelin D R 2003 *Phys. Rev. Lett.* **91** 157202
[5] Furdyna J K 1988 *J. Appl. Phys.* **64** R29
[6] Wolf S A, Awschalom D D, Buhrman R A, Daughton J M, Molnar S, Roukes M L, Chtchelkanova A Y and Treger D M 2001 *Science* **294** 1488
[7] Yuhas B D, Fakra S, Marcus M A and Yang P 2007 *Nano Lett.* **7** 905
[8] Seong H K, Kim J Y, Kim J J, Lee S C, Kim S R, Kim U, Park T E, Choi H J 2007 *Nano Lett.* **7** 3366
[9] Hger E and Osuch K 2003 *Europhys. Lett.* **63** 90
[10] Limpijumnong S and Van de Walle C G 2004 *Phys. Rev. B* **69** 035207

Journal of Physics: Conference Series **292** (2011) 012015 doi:10.1088/1742-6596/292/1/012015

One-pot synthesis of Fe-Co nanospheres by modified polyol process and their structural, magnetic studies

G Suresh[1], P Saravanan[2], D Rajan Babu[1]

[1]Advanced Materials Research Centre, Crystal Growth & Crystallography Division, School of Advanced Sciences, VIT University, Vellore-632 014, India.

[2]Advanced Magnetics Group, Defence Metallurgical Research Laboratory, Hyderabad-500 058, India.

drajanbabu@vit.ac.in

Abstract. Fe-Co nanoparticles stabilized by polyvinylpyrrolidone (PVP) have been prepared using modified polyol process. When 5 mmol of PVP was used, aggregated nanoparticles with irregular shapes were resulted; whereas on increasing the PVP concentration to 10 mmol, highly spherical nanoparticles were obtained. In the present work, the influence of non-ionic stabilizer viz. PVP on the synthesis of spherical nanoparticles and their structural, macromagnetic properties have been reported.

1. Introduction

Synthesis of magnetic nanoparticles is of vast interest because of their applications such as ultrahigh density data storage [1], magnetic resonance imaging [2], ferrofluids [3], magnetic hyperthermia treatment [4] and drug delivery [5]. Synthesizing magnetic nanoparticles with narrow size distribution is a significant task for most of the applications. A variety of chemical methods such as co-precipitation [6], thermal decomposition [7], polyol process [8,9] and sonochemical synthesis [10] are reported for synthesizing nanoparticles. Among all the methods, polyol process has been demonstrated as an excellent choice for synthesizing highly crystalline and monodispersed magnetic nanoparticles, as firstly reported by Micheal Figlarz et al. [11]. Among all the binary transition metal alloys, Fe-Co alloys are well known to possess a unique combination of high saturation magnetization, high Curie temperature, low magnetocrystalline anisotropy and good strength. They are ideally suited for applications requiring high flux densities [12]. The saturation magnetization of $Fe_{72}Co_{28}$ is 242 emu/g [13,14]. Synthesis of Fe-Co nanoparticles by polyol process is successful in controlling the size and morphology [8]. Fe-Co nanoparticles with a variety of morphologies such as spherical [8], dice [9], microcages [15], hollow spheres [16] and acicular [17] have been synthesized through chemical methods. When compared to other chemical synthesis techniques, polyol process is the best one for fine tuning the morphology of the synthesized nanoparticles [18]. Synthesizing spherical nanoparticles is a subject of intense interest for various technologically important soft magnetic applications, since the morphology is also one of the deciding parameters of coercivity [19]. Moreover, spherical particles do not introduce shape anisotropy. To avoid conflicts due to the shapes in the magnetic properties, we have synthesized spherical nanoparticles. We followed the method reported by Daisuke Kodama et al. [9]. They have reported the synthesis of Fe-Co nanocubes by varying refluxing temperature from

403 K to 470 K and they were able to synthesize nanocubes at 403 K. We modified the synthesis process performing the reaction at 470 K but varying the stabilizer concentration which had resulted in Fe-Co nanospheres. Then, structural and macromagnetic properties of as-synthesized Fe-Co nanospheres have been studied.

2. Experimental

Nanoparticles of Fe-Co were synthesized by the modified polyol process [9]. All the chemicals were obtained from commercial sources with ultra purity and used without further purification. Iron chloride tetrahydrate was procured from M/S Fluka and cobalt acetate tetrahydrate and other reagents were procured from M/S Merck chemicals, India. In a typical synthesis, 5 mmol of iron chloride tetrahydrate, 5 mmol of cobalt acetate tetrahydrate, 0.4 mol of sodium hydroxide and polyvinylpyrrolidone (PVP) were dissolved in 1 litre of ethylene glycol and purged with 99.99 % pure nitrogen gas (N_2 gas) for two hours. The solution was refluxed at boiling point for two hours under the nitrogen atmosphere. The formation of Fe-Co nanoparticles was witnessed by the colour change from dark red to black. The solution was cooled to room temperature and flocculent viz. ethanol was added to precipitate, and the supernatant was discarded. Using a strong magnet synthesized Fe-Co nanoparticles were separated. The resultant Fe-Co nanoparticles were washed thrice with ethanol and acetone to remove the reaction by-products and allowed to dry at ambient atmosphere. The above synthesis was performed using different PVP concentrations of 5 mmol, 10 mmol and the as-synthesized samples were named as S1 and S2 respectively. The typical reaction mechanism for the formation of Fe-Co nanoparticles is [20,21]:

$$FeCl_2.4H_2O + CoC_4H_6O_4.4H_2O + C_2H_6O_2 + NaOH \xrightarrow[\Delta]{} Fe\text{-}Co + C_4H_6O_2 + NaCl + CH_3COOH + H_2O$$

3. Results and Discussion

The main issue in synthesizing iron based metallic nanoparticles is oxidation. During the synthesis, the presence of trace amount of oxygen would aid the formation of ferrites [22]. Hence, in the present synthesis, the precursor solution was degassed with N_2 gas for two hours to eliminate the dissolved molecular oxygen and the whole reaction was carried out under nitrogen atmosphere. When the precursors were refluxed at boiling point the colour of the solution changed from dark to black which indicated the formation of Fe-Co nanoparticles. The colour of the synthesized nanoparticles may give a hint about the formation of Fe-Co nanoparticles. The oxide counterpart of Fe-Co is $CoFe_2O_4$, having a dark brown colour that was not evidenced during the synthesis [22,23]. We ended with black powder. X-ray diffraction (XRD) pattern of as-synthesized Fe-Co nanoparticles is shown in Figure 1(a) which apparently confirms the formation of disordered bcc phase, the peak can be easily assigned to (110) plane reflection of bcc Fe-Co alloy (JCPDS card number 44-1433). The composition of the as-synthesized nanoparticles was measured by atomic absorption spectroscopy (AAS) and was found to be 42.2% of iron and 57.7% of cobalt. The iron content in the resultant nanoparticles is always less than the charged [24]. For both the samples S1 and S2 iron and cobalt content is almost same. The functionalization of PVP was studied with Fourier transform infrared spectroscopy (FT-IR). As it can be seen from Figure 1(b), all the vibrations of the PVP are clearly noticed in both the samples (S1 and S2), confirming the functionalization of PVP over Fe-Co nanoparticles [25]. The slight deviation in the vibrations is due to the interaction of PVP molecules with the alloy nanoparticles and the observed vibrations are listed in table 1. Figure 2(a) and Figure 2(b) shows the agglomerated Fe-Co nanoparticles and spherical nanoparticles. Morphology and dimensions of the as-synthesized nanoparticles strongly depend on reaction conditions such as temperature, concentration of precursors, and the molar ratio between the repeating unit of stabilizer and precursors [26]. In the present work, for the synthesis of both the samples S1 and S2, all the conditions were constant except the stabilizer concentration. When 5 mmol of stabilizer was used, more surface of the nanoparticles is exposed to the solvent resulting in agglomeration to reduce the total surface area and the energy of the coated

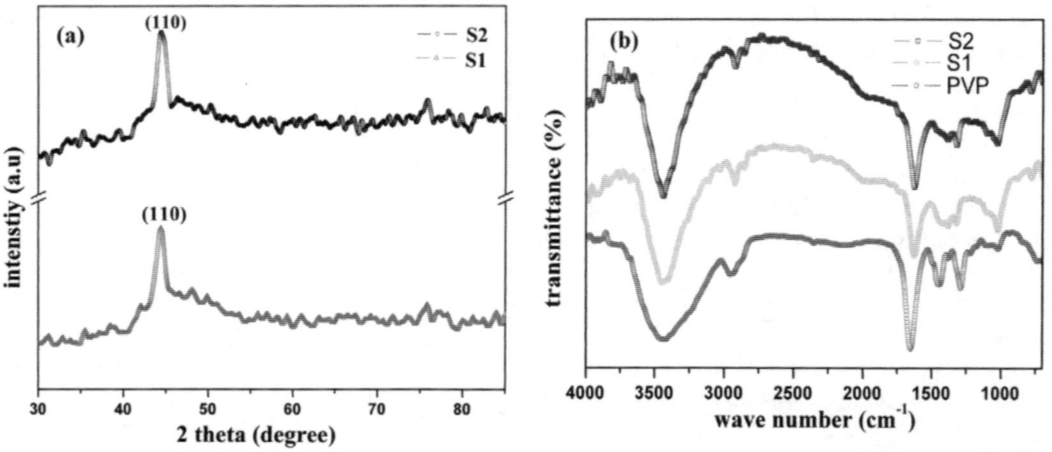

Figure1 (a). XRD pattern of Fe-Co nanoparticles functionalized with PVP (S1 and S2) **Figure1 (b).** FT-IR spectrum of PVP functionalized Fe-Co nanoparticles (S1 and S2)

surface. The strength of this process is proportional to the amount of the stabilizer with respect to any given amount of non-coated material. A larger amount of the stabilizer (10 mmol), allows for coating a larger surface, which becomes inactive and thus limits agglomeration [19]. If the synthesized nanoparticles are magnetic, the magnetic dipolar interactions also tend to agglomerate, which is possible because the particles are mobile in the reaction medium. The other possible reason for obtaining spherical shape at 10 mmol of PVP concentration could be the reaction time employed (two hours), as it is obvious that all the nucleated nanoparticles will transform into thermodynamically stable spherical shape, when they are heated for sufficiently longer time [18, 19, 27]. In total, the reaction temperature and higher concentration of stabilizer (10 mmol) facilitated the synthesis of spherical nanoparticles.

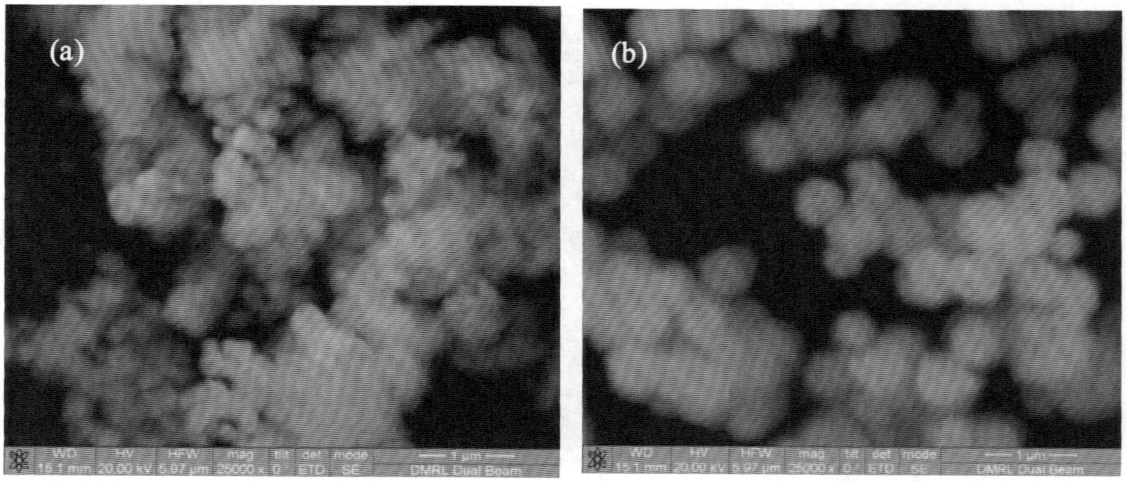

Figure 2. Scanning electron micrograph of Fe-Co nanoparticles (a) S1 and (b) S2

Table 1. FT-IR Bond vibration of the samples S1 and S2

Wave number (cm^{-1})	Bond vibration
3454	O-H
2940	C-H
1660	C=O
1482 1457	PVP bond absorption
1280	N→H-O complex

Table 2. Macromagnetic properties of the samples S1 and S2

Parameter	S1	S2
Coercivity (Oe)	93.4	158.4
Saturation magnetization Ms (emu/g)	141.8	104.9
Magnetic remanence (Mr) (emu/g)	0.099	0.140
Squareness (Mr/Ms)	0.041	0.077

The magnetic hysteresis measurements were carried out at room temperature in vibrating sample magnetometer (VSM) and the hysteresis loop is shown in Figure 3(a). The region highlighted in Figure 3(a) is magnified and shown as Figure 3(b) to demonstrate the ferromagnetic nature of S1 and S2. The macromagnetic properties of the samples are listed in table 2. In the present study, coercivity values have been determined as 93.4 and 158.4 Oe for the samples S1 and S2 respectively. The increase in coercivity in S2 may be due to the size dependent property of the synthesized nanoparticles. By increasing the stabilizer concentration, the coercivity increases by 65 Oe. It is well known that in single domain particles, the coercivity tends to reduce with reduction in the particle size [25]; when the particle size is reduced less than 10 nm, it apparently falls on the super paramagnetic limit [29,30].

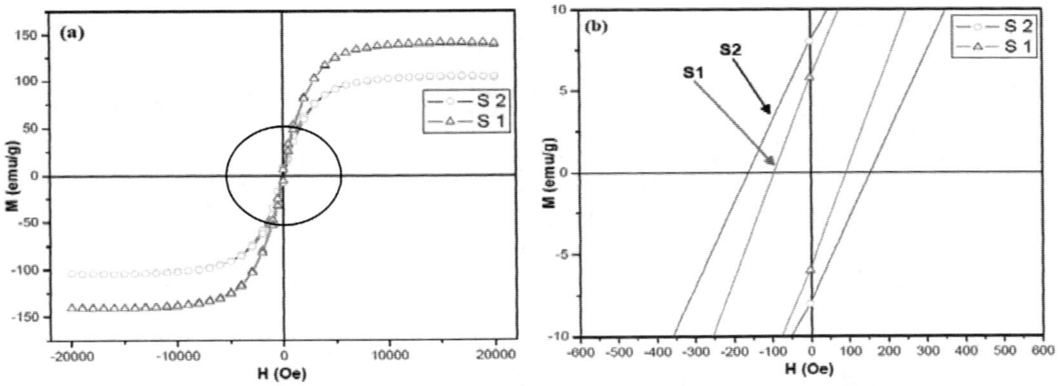

Figure 3. **(a)** Hysteresis loops of Fe-Co nanoparticles (S1 and S2). **(b)** Magnified hysteresis loop of S1 and S2

The observed saturation magnetization values for the samples S1 and S2 are 141.8 and 104.9 emu/g respectively. These values are significantly reduced as compared to that of their bulk values $Fe_{50}Co_{50}$ (211 emu/g) [15]. The reasons for the reduction of saturation magnetization are decrease in iron content in the samples i.e. $Fe_{42.2}Co_{57.7}$ than charged in the reaction $Fe_{50}Co_{50}$, lack of crystallinity of the as-synthesized nanoparticles (disordered bcc phase) and the stabilizer masking. Annealing the samples in inert atmosphere can greatly enhance the saturation magnetization. Currently we are exploring the possible route to synthesize colloidal Fe-Co nanoparticles for sophisticated soft magnetic applications.

4. Conclusions

In summary, Fe-Co nanospheres have been synthesized using modified polyol process and their structural, macromagnetic properties have been studied. The X-ray diffraction pattern confirms the formation of disordered bcc phase; FT-IR studies show the functionalization of PVP on Fe-Co nanospheres. The increased PVP concentration from 5 mmol to 10 mmol effectively yields spherical nanoparticles, which may be due to the larger amount of PVP molecules which efficiently stabilizes the nucleated nanoparticles. Adversely, the increased concentration of PVP also contributes to decrease the saturation magnetization from 141.8 to 104.9 emu/g, other probable reason for the reduction in saturation magnetization may be the lack of crystallinity and the compositional effects of the synthesized nanoparticles. But the major advantage of increased PVP concentration is that agglomeration is reduced and spherical nanoparticles are formed.

Acknowledgments
One of the authors, G. Suresh, wishes to express his sincere gratitude to the Management, VIT University, Vellore-632 014, India for Research Associateship and financial assistance to carry out his research work. He is also grateful to Prof. Dr. K. Sathiyanarayanan, Director, SAS and Dr. R. Ezhil Vizhi, VIT University, Vellore-632 014, India for their continuous support and encouragement.

References
[1] Sun S et al. 2000 *Science* **287** 1989.
[2] Seo W S et al. 2006 *Nat.Mater.* **5** 971.
[3] Zahn M 2001 *J.Nanopart.Res.* **3** 73.
[4] Hergt R et al. 1998 *IEEE Trans. on. Magn.* **34** 3745.
[5] Chen. F H et al. 2008 *Nanotechnology* **19** 165103.
[6] Shin S J et al. 2007 *Current Applied Physics* **7** 404.
[7] Desvaux C et al. 2005 *Nat.Mat.* **4** 750.
[8] Chaubey G S et al. 2007 *J.Am.Chem.Soc.* **129** 7214.
[9] Kodama D et al. 2006 *Adv.Mater.* **18** 3154.
[10] Li Q et al. 2003 *New J.Chem.* **27** 1194.
[11] Figlarz M et al. 1985 *US Patent* 4539041.
[12] Sundar R S and Deevi S C 2005 *Int. Mater. Rev.* **50** 157.
[13] Bardos D I 1969 *J.Appl.Phys.* **40** 1371.
[14] Kuhrt C and Schultz L 1993 *J.Appl.Phys.* **73** 6588.
[15] Wei X W et al. 2008 *Chem. Mater.* **20** 6248.
[16] Huang J et al. 2007 *Nanotechnology* **18** 415603.
[17] Pozas R et al. 2007 *Nanotechnology* **18** 205601.
[18] Puntes V F et al. 2002 *Top.Catal.* **19** 145.
[19] Prozorov T et al. 1999 *Thin Solid Films* **340** 189.
[20] Sun J et al. 2005 *Mater. Lett.* **59** 3933.
[21] Larcher D and Patrice R 2000 *J.Solid StateChem.* **154** 405.
[22] Bhattacharyya S et al. 2005 *Chem.Commun.* 4818.
[23] Kim Y et al. 2003 *Physica B:Condensed Matter* **337** 42.
[24] Chen M and Nikles D E 2001 *Nano Lett.* **2** 211.
[25] Zhang Z et al. 1996 *J.Sol.State.Chem.* **121** 105.
[26] Sun Y and Xia Y 2002 *Science* **298** 2176.
[27] Zhou Q F et al. 2002 *J. Mater.Chem.* **12** 384.
[28] Leslie-Pelecky D L and Rieke R D 1996 *Chem.Mater.* **8** 1770.
[29] Hesani M et al. 2010 *Solid State Commun.* **150** 594.
[30] Huber D L 2005 *Small* **1** 482.

Heat treatment of mechano-chemically produced $BaFe_{12}O_{19}$ / Fe_3O_4 magnetic nano-composites

M J Molaei[1], A Ataie and S Raygan

School of Metallurgy and Materials Engineering, College of Engineering, University of Tehran, P.O. Box: 14395-553, Tehran, Iran

e-mail: mjmolaee@ut.ac.ir

Abstract. Barium hexaferrite ($BaFe_{12}O_{19}$) magnetic powder was synthesized via ceramic route by mixing of Fe_2O_3 and $BaCO_3$ and subsequent calcination of the mixed powder in air at 1150°C. The synthesized barium hexaferrite was mixed with graphite to be treated mechano-chemically in high energy ball mill in air atmosphere for various times. Subsequently, the milled samples were subjected to high temperature in argon atmosphere. XRD patterns of the milled samples revealed that graphite partially reduced barium hexaferrite to Fe_3O_4 after 15 hours of milling, resulted in a magnetic nano-composite of $BaFe_{12}O_{19}/Fe_3O_4$. Extending milling time to 80 hours promoted the reduction process to FeO. Heat treatment of the milled samples resulted in different magnetic composites. Iron was seen in XRD pattern of the 40 hours milled sample that was heat treated at 680°C. SEM images showed that after heat treatment, fine particles in agglomerates grew and larger particles formed. As a result, mean particle size increased due to the heat treatment of the milled powder.

Keywords: Barium hexaferrite; magnetic nano-composite; ball milling; partial reduction, heat treatment

1. Introduction

Hexagonal ferrites, especially barium hexaferrite ($BaFe_{12}O_{19}$) have been used for different applications such as permanent magnets, magnetic recording media, magneto-optic materials and microwave filters. These materials exhibit high saturation magnetization and high values of coercivity as well. High Curie temperature and chemical stability are other advantages [1, 2].

Due to recent interest for production and application of hexaferrite nanoparticles, many researchers have employed different methods such as co-precipitation, sol–gel combustion, micro-emulsion, hydrothermal and self-propagating high temperature synthesis to produce this type of nano-materials [3-6].

[1] Corresponding author

Low values of coercivity can be achieved by heat treatment of hexaferrite materials under carbon containing gases, nitrogen and hydrogen or a combination of these gases at high temperatures. Lower coercivity makes these materials suitable for magnetic recording media such as hard disks, video tapes, and etc. where a lower coercivity is necessary for easier data re-writing [7-10]. Here, the effect of milling of barium hexaferrite and graphite mixture in a high energy ball mill and subsequent heat treatment on magnetic properties has been investigated.

2. Experimental procedure

Hematite (>99% purity) and barium carbonate (>99% purity) were mixed in a jar mill and calcinated at 1150 °C for 2 hours. The molar ratio of $Fe_2O_3/BaCO_3$ was 5.5/1. The synthesized barium ferrite was mixed with graphite. The molar ratio of C/O was 1.1 in the mixture. The powder blend milled for different times in a high energy planetary ball mill with ball to powder mass ratio and rotating speed of 35 and 300 rpm, respectively. Heat treatment of the samples carried out under argon atmosphere in an electrical tube furnace with heating rate of 10 °C/min and dwell time of 10 minutes at the desired temperature. Then, the samples were furnace-cooled under argon atmosphere. Phase identification was done with a Philips PW-1730 XRD machine with Co-Kα radiation. Powder crystallite size was calculated according to Williamson-Hall method [11] (see appendix). DTA analysis (Differential Thermal Analysis) performed in a NETZSCH STA 409 PC/PG calorimeter with Al_2O_3 crucible and heating rate of 10 °C/min under flowing argon atmosphere. Magnetic properties measurements were done by a LakeShore Model 7307 VSM. Philips XL30 SEM was used to study particle size and morphology of the powder.

3. Results and discussions

The XRD patterns of milled barium hexaferrite and graphite mixtures are shown in figure 1. It can be seen that during milling, Fe_3O_4 and then FeO phases have been produced from $BaFe_{12}O_{19}$. In fact $BaO \cdot 6(Fe_2O_3)$ is reduced to lower iron oxides in high energy ball milling. The reduction process makes it possible to obtain a $BaFe_{12}O_{19}/Fe_3O_4$ magnetic composite during milling. The particles of barium hexaferrite which is a brittle ceramic material fractures repeatedly in the milling medium to reach the nanoscale. The soft graphite which surrounds these particles makes a nano-reactor [12] for the reduction process of barium hexaferrite to lower iron oxides.

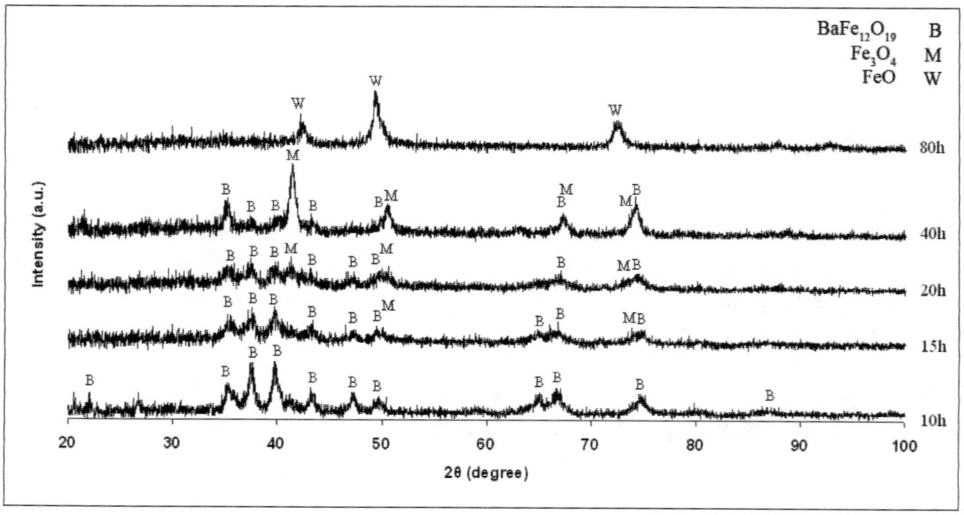

Figure 1. XRD patterns of barium hexaferrite and graphite mixtures milled under air atmosphere for different times; 10, 15, 20, 40, and 80 hours.

After 15 hours of milling, the intensity of Fe_3O_4 peaks in the XRD patterns increases and that of $BaFe_{12}O_{19}$ decreases with milling time. Long milling times up to 80 hours results in an improvement in reduction process and appearance of FeO peaks in the XRD patterns. XRD peak broadenings shows that the crystallite size of barium hexaferrite is 41 and 14 nm for 10 and 20 hours milled samples, respectively. This parameter is 7 nm for FeO in 80 hours milled sample. Therefore, increasing milling time results in decreasing crystallite size of barium hexaferrite in the milled samples.

Figure 2 represents DTA curves of un-milled and 40 hours milled mixtures of barium hexaferrite and graphite. The endothermic peak at 992 °C in the curve of un-milled sample which relates to the reduction of barium hexaferrite to iron and other Ba containing compound(s) decreases to 748 °C after 40 hours milling of the mixture. This decrease in the peak temperature is a result of accumulation of energy and inducing crystalline defects into the powder particles due to the milling.

Figure 3 represents XRD patterns of un-milled and 40 hours milled samples that are heat treated at 680 °C. These patterns illustrate that for un-milled sample, reduction process proceeds to FeO due to the heat treatment. Heat treatment of the 40 hours milled sample resulting in a new composite consisting of Fe, Fe_3O_4, $BaFe_{12}O_{19}$, FeO, C, $BaFe_2O_4$ and $Ba_3Fe_2O_6$ phases.

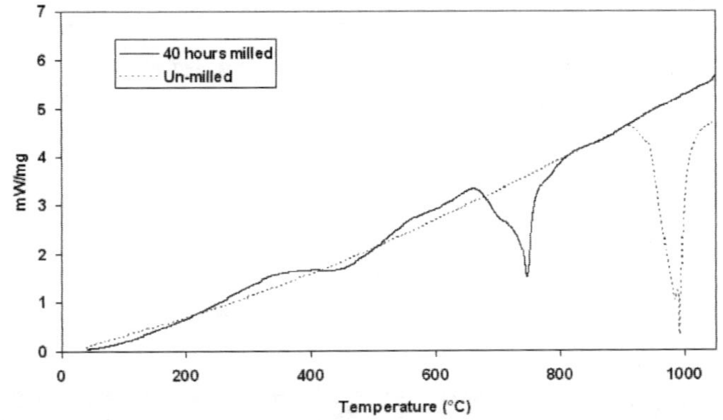

Figure 2. DTA diagrams of barium hexaferrite and graphite mixtures for un-milled and 40 hours milled samples.

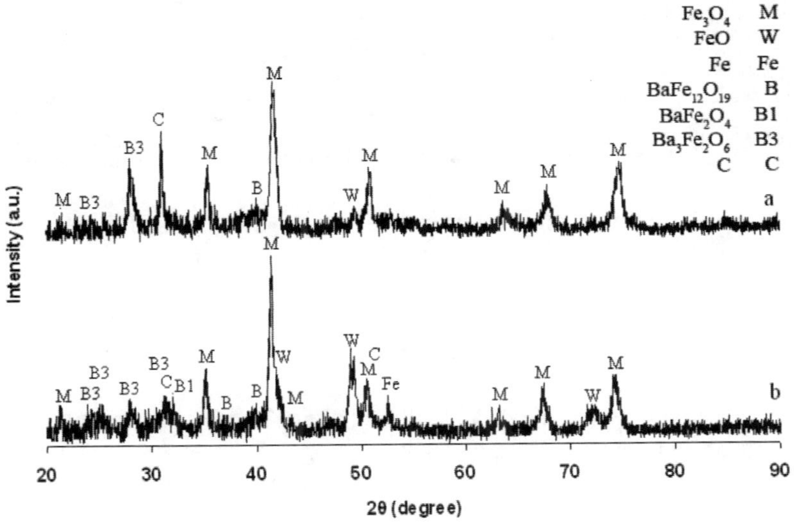

Figure 3. XRD patterns of heat treated samples at 680 °C; (a) un-milled and (b) 40 hours milled.

Figure 4 shows the hysteresis loops of as-produced and heat treated samples. By heat treating the 40 hours milled sample, coercivity of the initial hard magnetic material decreases because of the formation of a new magnetic composite of $BaFe_{12}O_{19}$ (hard magnetic material), Fe (soft magnetic material), Fe_3O_4 and some other non-magnetic phases. The intrinsic coercivity, H_{ci}, decreases from 2020 Oe for as-produced powder to 477 Oe for milled and heat treated powder.

Figure 5 shows SEM images of 40 hours milled samples before and after heat treatment and also 80 hours milled powder. Figures 5(a) and 5(b) show that each larger agglomerate consists of fine particles. The mean particle size for 40 and 80 hours milled samples is 94 and 56 nm, respectively. It can be seen in figure 5(c) that after heat treatment, fine particles in agglomerates of milled powder adhered to each other and grew to form larger particles.

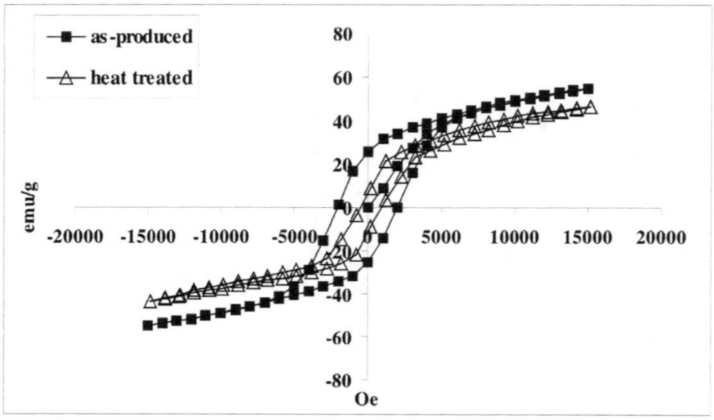

Figure 4. Hysteresis loops of as-produced and heat treated samples.

Figure 5. SEM images of barium hexaferrite and graphite mixture; (a) milled for 40 hours, (b) milled for 80 hours and (c) milled for 40 hours and then heat treated at 680 °C.

4. Conclusions

According to the results of the present research, the following conclusions were drawn.

1. By milling of $BaFe_{12}O_{19}$ and graphite, a nano-crystalline magnetic composite consisting of $BaFe_{12}O_{19}$ and Fe_3O_4 was formed after 15 hours. Extending milling time resulted in increasing the amount of Fe_3O_4 in this composite.
2. FeO phase observed in the 80 hours milled sample due to the reduction process. The crystallite size of this phase was 7 nm at this stage.

3. Heat treatment of the un-milled mixture for 10 minutes at 680 °C promoted the reduction process to FeO phase. This heat treatment for 40 hours milled sample promoted the reduction process up to iron and made a magnetic composite consisting of $Fe/Fe_3O_4/BaFe_{12}O_{19}$ beside other phases. The coercivity decreased as a result of heat treatment of the milled sample.
4. The particle size of the milled samples reached to 94 and 56 nm after 40 and 80 hours of milling, respectively. Furthermore, heat treatment resulted in particle size increments.

Acknowledgment
The authors would like to appreciate School of Metallurgy and Materials Engineering, University of Tehran and Iran Nanotechnology Initiative Council for financial support of this project. In addition, M.J. Molaei would like to thank Iranian National Elites Foundation for financial support of this work.

Appendix

Williamson-Hall [11] method for crystallite size measurements relies on the principle that the contribution of size broadening, β_L, and strain broadening, β_e, of the XRD peaks width varies with Bragg angle as follow:

$$\beta_L = k\lambda/L \cos\theta$$

$$\beta_e = C_\varepsilon \tan\theta$$

The contribution of β_L varies with $1/\cos\theta$ and that of β_e varies with $\tan\theta$. Total contribution of size broadening and strain broadening will be:

$$\beta_{tot} = \beta_e + \beta_L = k\lambda/L \cos\theta + C_\varepsilon \tan\theta$$

Or

$$\beta_{tot} \cos\theta = C_\varepsilon \sin\theta + k\lambda/L$$

By plotting $\beta_{tot}\cos\theta$ versus $\sin\theta$ for different peaks, C_ε and $k\lambda/L$ can be obtained from a linear fit as the slope and intercept, respectively.

References

[1] Campbell P 1994 *Permanent Magnet Materials and Their Application* (Cambridge: Cambridge University Press)
[2] Topal U, Ozkan H and Sozeri H 2004 *J. Magn. Magn. Mater.* **284** 416–22
[3] Pillai V, Kumar P and Shah D O 1992 *J. Magn. Magn. Mater.* **116** L299-L304
[4] Mali A and Ataie A 2004 *Ceram. Int.* **30** 1979–83
[5] Li K, Gu H and Wei Q 2004 *China Particuology* **2-1** 41-43
[6] Nikkhah-Moshaie R, Seyyed Ebrahimi S A and Ataie A 2008 *Mat. Sci. Eng.* A **473** 244–48
[7] Ataie A, Harris I R and Ponton C B 1999 *United States Patent,* Patent Number: 5858265
[8] Ataie A, Ponton C B and Harris I R 1996 *J. Mater. Sci.* **31** 5521-27
[9] Seyyed Ebrahimi S A, Ponton C B, Harris I R and Kianvash A 1999 *J. Mater. Sci.* **34** 35-43
[10] Seyyed Ebrahimi S A, Ponton C B and Harris I R and Kianvash A 1999 *J. Mater. Sci.* **34** 45-52
[11] Williamson G K and Hall W H 1953 *Acta Metallurgica* **1(1)** 22-31
[12] Kashiwaya Y and Ishii K 2004 *ISIJ Int.* **44-12** 1981-1990

AUTHOR INDEX

Institute of Physics
Dirac House, Temple Back
Bristol BS1 6BE UK